MANAGING THE
OUTER CONTINENTAL SHELF LANDS

Oceans of Controversy

R. Scott Farrow

with

James M. Broadus
Thomas A. Grigalunas
Porter Hoagland III
James J. Opaluch

Taylor & Francis

New York • Bristol, PA • Washington D.C. • London

USA	Publishing Office:	Taylor & Francis New York Inc.
		79 Madison Ave., New York, NY 10016-7892
	Sales Office:	Taylor & Francis Inc.
		1900 Frost Road, Bristol, Pa 19007
UK		Taylor & Francis Ltd.
		4 John St., London WC1N 2ET

Managing the Outer Continental Shelf Lands: Oceans of Controversy

First published 1990
Printed in the United States of America

Library of Congress Cataloging in Publication Data

Farrow, R. Scott
    Managing the Outer Continental Shelf Lands: oceans of controversy
/R. Scott Farrow with James Broadus . . . [et al.].
        p.   cm.
    Includes bibliographical references.
    ISBN 0-8448-1657-4. -- ISBN 0-8448-1658-2 (pbk.)
    1. Mineral resources in submerged lands—United States.
    2. Continental shelf—United States.   1. Titles.
    TN264.F37   1990
    333.8'5'09162--dc20   89-28048
                                                                             CIP

To Don (1947–1976)
In recognition of his struggle to find the balance
among his love of nature
his fight for social justice
and his own pursuit of happiness

Contents

v

Part III. Emerging Policy Issues

Preface

The hot sands of Southern California beaches and the sets of Pacific waves were delights of my childhood. Later on I lived by the Atlantic and would feel the difference between the oceans as I walked along the beach and fished from the shore. I was shocked in 1969 by the Santa Barbara oil spill and numbed in 1989 by the Exxon Valdez oil spill. My consciousness a child of the sixties, I thought we were using resources too fast and worried about limits to growth. I worked as a migrant laborer, lived in a tepee, unloaded sand from boxcars, and joined with friends to create an organic farm devoted to self-sufficiency. In time I felt and learned that self-sufficiency is a relative concept and natural resource policy for a society was more intricate than it appeared at first glance.

My concerns about pollution and the rate of resource use did not disappear over the years; instead they became embedded in an increasingly intricate web of cause and effect, of personal decisions, and the way in which our society reaches public decisions. I became familiar with the aphorism that not to decide is, in fact, to decide, and that to pursue one course of action results in a cost measured by opportunities that are passed by. I observed the struggle that goes on over wealth; and I saw in particular how decisions about managing the Outer Continental Shelf lands can transfer funds to the government, to the states, or to the oil companies. I also observed the concern of coastal residents who feared for lost property values and diminished enjoyment of the beaches I had walked along. I saw that the question of property ownership becomes more complex when one's actions affect many others and because ownership of ocean resources is a peculiar mixture of coastal state ownership, federal government ownership, and private ownership. In the meantime, I've seen oil rigs from shore, landed on them from the air, and been over a mile deep in onshore mines as I investigated how companies that extract natural resources make decisions.

As I sought answers to my personal questions concerning the problems of pollution, whether we use resources too fast, and who benefits from the decisions, I found the dangers of seeking to impose my view on others. History has shown us too often that people who know they are right are often wrong when they impose a plan of action on a society. I found that the process by which a society reaches a decision is critical.

Through work in the Carter and Reagan administrations and as an academic, I found that logic and data provide a partial answer to the questions that plague me and that plague the management of the Outer Continental Shelf. Feelings supply a different type of partial answer to these questions. This book is an attempt to use feelings about what is right and wrong, about what are large issues and small issues, to motivate studying theories and data useful to the management of the Outer Continental Shelf lands. The answers are not final, but decisions are made daily in the face of uncertainty, and so I and my coauthors offer this book as our thoughts on the fact that managing the Outer Continental Shelf does lead to oceans of controversy, but that controversy, like crisis, is both a danger and an opportunity.

I am fortunate to have two chapters contributed by outstanding researchers familiar with OCS management. Chapter 4 on the environment is written by Thomas A. Grigalunas and James J. Opaluch of the University of Rhode Island. James M. Broadus and Porter Hoagland III of the Woods Hole Oceanographic Institution wrote Chapter 7 on nonfuel marine minerals. The expertise of these authors (short biographies follow the index) is matched only by their congeniality as coauthors.

Many other people have contributed to the preparation of this book. Without implication, I have benefited greatly from discussions with Marshall Rose, Ted Heintz, Don Rosenthal, Bob Hahn, Jeff Krautkraemer, Dennis Epple, Mark Kamlet, Steve Garber, Bill Watson, Steve Polasky, Joe Swierzbinsky, Mike Toman, Charles Mason, Gregory M. Duncan, Andrew Solow, and Mary Vavrina. Those who developed the data in the five-year plan of the Minerals Management Service as well as many others in the field have helped by their work, criticism and descriptions. I also thank William Bettenberg and Carolita Kallaur for their insightful interviews. Patricia Malinowski provided important editorial advice while Teresa Morris and Ellen Gately patiently typed portions of the manuscript. On a more personal basis, the enthusiasm, proenvironmental views, and emotional support of my wife, Elaine A. King, have made this project not only feasible but also much more enjoyable.

My coauthors and I also appreciated support from the institutions that allowed us to focus our research efforts on this topic. They are the Marine Policy Center at the Woods Hole Oceanographic Institution, the J. N. Pew, Jr. Charitable Trust, the U.S. Department of the Interior, the Department of Engineering and Public Policy and the School of Urban and Public Affairs at Carnegie Mellon University, the National Sea Grant College Program of NOAA, U.S. Department of Commerce, the University of Rhode Island, contribution number 2522 of the College of Resource Development, with support from the Rhode Island Agricultural Experiment Station, and the Environment and Policy Institute of the East-West Center.

Scott Farrow

The School of Urban and Public Affairs
Carnegie Mellon University
Pittsburgh, Pennsylvania, and
the Marine Policy Center
The Woods Hole Oceanographic Institution
Woods Hole, Massachusetts

Part One

Management Purpose and Practice

The three parts of *Managing the Outer Continental Shelf Lands: Oceans of Controversy* describe, analyze, and predict policy issues for the management of the Outer Continental Shelf (OCS). Each part reflects the view that good policy analysis and good policy require both knowledge about institutions and skill in analysis. Analysis without institutional context is like a language without people—it can exist but it is seldom useful—whereas institutions unaffected by analysis are prey to being blindsided by the future.

The two chapters in Part One describe current management practice—the social and technical context of OCS management decisions and the goals of the institutions involved in OCS management. These chapters lay the foundation for what exists and for what should exist as seen from the vantage points of Congress, of the Department of the Interior, and of the states, local governments, and interest groups.

Chapter 1

The Practice of Outer Continental Shelf Management

Many people in many areas resist exploration and production of energy and minerals from beneath the ocean because of environmental and personal costs that are perceived to be too high. North Carolina has angrily resisted drilling offshore near Cape Hatteras. Massachusetts is up in arms about drilling on or near the fishing grounds of the Georges Bank. California consistently blocks efforts to expand offshore oil and gas activity near its population centers or its scenic areas. Louisiana worries about the loss of wetlands associated with offshore energy. Florida seeks to protect rare coral banks, and Alaska worries about its fisheries and tourism.

People who support energy production from the Outer Continental Shelf (OCS) lands have a different perspective and different arguments. They state that energy production is vital to support the lifestyle choices that people make every day. Individuals drive personal cars an average of 9,000 miles per year, using over 500 gallons of gasoline per passenger car. Over 80 million households spend an average of approximately $1,000 per year for household energy use. Because oil and gas once used is gone forever, new sources must constantly be found. Proponents of offshore energy exploration and production argue that offshore sources are one of the least-cost ways of meeting the future demand for energy.

The arguments for each side in the controversy include many types of costs, costs that vary in different parts of the country. Even when there is agreement on the type of costs, the costs themselves are very uncertain, as are the potential quantities of energy resources and the price at which they will sell. Yet decisions about offshore energy resources are made in the oil companies, in local and state governments, in the federal government, and in the environmental groups. Decisions in each of these organizations are intertwined, but all of them intersect in the Minerals Management Service of the U.S. Department of the Interior.

Officially, the Minerals Management Service is charged with the management of federal offshore lands—the lands beyond the three nautical miles claimed by most states to the 200-mile limit of the exclusive economic zone. These federally managed lands cover approximately 1.3 million square miles, equivalent to about one-third of the on-shore land mass of the United States. These lands directly abut the 23 coastal states, three foreign countries (Canada, Mexico, and Russia), and the international high seas.

Vast wealth as well as large areas are being managed. In the 34 year period starting in 1954 and extending to 1987, private oil and gas companies paid into the federal government $53 billion in pre-exploration (bonus) payments for the right to explore and, if fortunate, to extract oil and gas from these lands. In 1988 oil and gas production from the lands of the OCS amounted to 12 percent of domestically produced petroleum and 26 percent of domestically produced natural gas. It is no wonder that controversy exists over such wealth and the variety of other uses of the OCS.

This chapter introduces the practice of Outer Continental Shelf management. The first section discusses the current framework for leasing individual claims and the larger context of multiple uses of the OCS. The second section surveys the history of policies for land disposal and withdrawal. The final section describes the areas of uncertainty and surveys knowledge about geology, technology, ecology, politics, and economics that shape and constrain the current practice of Outer Continental Shelf management.

THE DUAL MANAGEMENT PROBLEM

Managers of the Outer Continental Shelf must attend to both the details of specific wells, leases, and animal species as well as the broad trends of national energy use, aggregate fisheries production, and maritime uses of the ocean. The dual management problem for the OCS is the management of the details, the micro-management problem, as well as the management of the broad trends, the macro-management problem. This section surveys some of the micro and macro management problems of the Outer Continental Shelf.

The micro-management problem of the Minerals Management Service is based on the details of the Outer Continental Shelf Lands Act and the delegation of authority to the Secretary of the Interior. Fundamentally, the act requires that offshore claims, called tracts or blocks, be leased through an auction process and that a royalty rate of at least 12 1/2 percent of gross income be charged the company and paid to the U.S. government.

The current framework, which is elaborated on throughout this book, is that an approximate schedule of auctions for five years is developed by the Minerals Management Service and then approved by Congress. Each proposed auction in the five-year plan covers a large area, say the central Gulf of Mexico. Extensive planning and consultation occurs before the number of tracts to be offered is finalized and before the actual auction date. As many as several thousand or as few as several hundred tracts may be offered at a particular auction.

If a company wishes to compete for the right to explore one of these tracts, they conduct surface exploration (they cannot drill) prior to the auction and decide if they wish to submit a bid. In most cases, this is a bonus bid. A bonus bid is a one-time payment if the company is the highest bidder for that particular tract in the auction. The companies submit sealed bids for the tracts in which they are interested. These bids are opened publicly to determine the highest bidder for a tract. This highest bid is then compared to regulations that establish the minimum for accepting bids; that minimum is typically $25 per acre. Most tracts exceed 5,000 acres.

In many cases, the high bid must also exceed a measure of the tract's value as computed by the government. If the high bid passes these bid adequacy criteria, the company receives the right to explore the area for a fixed period of time as specified prior to the auction. The primary lease term is typically five years in shallow water and 10 years in deeper water. At the end of the primary lease term, the land reverts to the federal government unless the company is actively drilling or producing on the tract, or unless a suspension of operations is received from the government. If the company is producing after the primary term, the lease is said to be in its secondary term, which occurs as long as production continues; a common length of time is 20 years. Finally, when production stops, the current regulations require that the production platform from which the oil and gas was produced be removed.

This summary of the micro-management problem is replete with contentious policy issues. What exactly are the current bid adequacy rules? What determines the amount that is bid for a tract? What is the effect of exploration on the value of tracts? These questions as well as many others are investigated in the policy analysis chapters of Part Two.

The macro-management problem of the Minerals Management Service places the details of management in the broader context of competing uses of the ocean, national energy policy, and environmental issues. One view of the macro-management problem can be seen by considering the extent and value of the many uses and resources of the Outer Continental Shelf.

One cause of debate about OCS management relates to the uncertainty of the quantity and value of its resources. Table 1-1 presents data about the area, estimated resources, and value in the four major regions of the OCS: the Gulf of Mexico, the Pacific region, Alaska, and the Atlantic. One estimate of the value of all future federal bonus and royalty revenue (if evaluated in 1981 dollars and at price and cost conditions prevailing at that time) amounts to over $600 billion for oil and gas alone.[1] These estimates do not include the value of other potential resources, such as special types of sand, phosphate and manganese crusts, and polymetallic sulphides that are expected to be mined from the OCS lands in the future. In contrast, Department of the Interior estimates of private profits as well as bonus and royalty revenues remaining in unleased fields are only $19 billion as of 1987. This large discrepancy is due to a variety of reasons including the change in price from 1981 to 1987, disappointing exploration results off-

Table 1-1
Estimated Value, Resources, and Area in the Four Regions of the OCS

	REGIONS				
	Gulf of Mexico	Pacific	Alaska	Atlantic	Total
Value (billion $)					
Boskin et al. (1981)					672.8
MMS (1987 low price)	16.8	1.8	—	.5	19.1
Resources[a]	9.8	2.0	3.1	2.3	17.2
Area (000 sq. mi.)	205.2	190.5	772.2	231.6	1,399.5
Percent	15	14	55	16	100

[a] In billions of barrels of oil equivalent, undiscovered unleased economically recoverable, 1987.
Sources: M. Boskin et al., "New Estimates of the Value of Federal Mineral Rights and Land," *American Economic Review* (December 1985). Minerals Management Service, *Federal Offshore Statistics*, 1987; Department of the Interior, *Five Year Leasing Program, Mid 1987 to Mid 1992*, July 1987.

shore of Alaska and in the North Atlantic region, and different approaches for mea-suring value.[2]

The region-specific nature of some controversies is also evident in Table 1-1; line two indicates that the $19-billion-estimate by the Department of the Interior is dispro-portionately located in the Gulf of Mexico and off the Pacific Coast, primarily along the California coast. As of 1987 the Minerals Management Service did not believe that there were any economically recoverable resources to be leased in Alaska at that time, though Alaska accounts for over 50 percent of the potentially exploitable area on the OCS as indicated by line 5 in Table 1-1.

These disparities in estimation and within regions are an indication of a fundamental macro-management problem for the OCS. That problem is the large degree of uncer-tainty due to different levels of exploration in different areas and to the changing nature of what is economically recoverable as prices, costs, and technology change over time. For instance, the resource and value estimates of the Minerals Management Service are very sensitive to changes in the price. The Minerals Management Service estimated that the present value of the OCS would increase by a factor of four if the starting oil price were to double. Similar results were obtained for the estimates in line 1. Regardless of the particular estimates used, the lands of the Outer Continental Shelf currently com-prise one of the nation's most valuable publicly held assets.

A second problem in the macro-management of the Outer Continental Shelf is in-dicated by the variety of ways in which the OCS is already productive. OCS lands and the water above them support oil and gas production, fisheries, and shipping activities, and indirectly affect coastal uses such as tourism. Table 1-2 presents data for oil and gas production and value, fisheries, and tourism figures for the major OCS regions. Examples of the predominant role of the Gulf of Mexico in OCS production is illustrated

Table 1-2
Uses of the Outer Continental Shelf[a]

	REGION				
	Gulf of Mexico	Pacific	Alaska	Atlantic	Total
Square miles leased (1986)	50,625	3,906	7,656	3,594	65,781
Percent	77	6	12	5	100
Output (1985)					
Oil (million bbl)	359	29	—	—	389
Gas (tril. ft^3)	3,951	49	—	—	4,001
Value (1985, million $)					
Oil and gas production	21,211	779	—	—	21,991
Royalty	3,490	148	—	—	3,638
Bonus and rental	1,531	2	11	2	1,547
Bonus and rental (1954 through 1985)	40,339	3,985	5,893	2,869	53,086
Fisheries (1983, million $)	607	302	564	844	2,317
Tourism (million $)	16,427	4,568	3,444	31,860	53,199

[a] All data exclude Hawaii; tourism also excludes Oregon and Washington.
Sources: Minerals Management Service, *Federal Offshore Statistics*, 1985, OCS Report MMS 87-0008; Department of the Interior, *Five Year Leasing Program, Mid 1987 to Mid 1992*, (July 1987).

in lines one through three of the data. This region accounts for almost three-quarters of the leases and an even larger proportion of production. By comparison, Alaska and the Atlantic are more prospective regions, that only became worth investing in during the rapid price increases of the 1970s and early 1980s. The subsequent decline in price and relative lack of successful exploration in these areas leave the Gulf of Mexico and California as the only major proven producing regions on the OCS as of 1989.

In 1985 production of oil from the OCS amounted to 13 percent of domestic oil production, whereas the production of gas from the OCS amounted to 25 percent of domestic gas production. The value of this oil and gas production amounted to almost $22 billion, with the government receiving $3.6 billion in royalty payments and additional income from taxes (Table 1-2). For the right to explore and produce in these areas (including areas that to date have been unproductive), oil companies have paid over $81 billion in bonus bids and royalty payments since 1954. The majority of these payments go directly into the treasury for general spending, though approximately $1 billion per year is earmarked for historic preservation and a land and water conservation fund. Since 1978 a share of the revenues has also been earmarked for distribution to the coastal states.

The 1985 OCS energy production value of $22 billion is complemented by the commercial fishery production value of $2.3 billion and estimates of the tourism trade equal to $53.2 billion. These data indicate that multiple-use management of the OCS is possible. The clearest example of multiple-use management can be found in the Gulf of Mexico area where income generated by oil and gas production is the largest and revenue from the commercial fishing and tourism industries is second largest.

Only the most visible activities associated with the OCS, such as oil and gas production, fishing, tourism, and other coastal land use, are included in Table 1-2. Other activities such as maritime transportation, defense uses, marine habitat, and scientific uses are not directly included.

Several macro-management issues are investigated in detail in the policy analysis chapters of Part Two and in Chapter 7 on the non-fuel minerals of the Outer Continental Shelf. Some of the macro-management issues addressed in later chapters include a policy to allow competing users of the OCS to participate in the OCS auction process, whether or not the outcome of the five-year plan of the Minerals Management Service does incorporate nonoil and gas uses of the Outer Continental Shelf, and the potential of nonfuel minerals to represent an expanded use of the OCS lands.

The micro- and macro-management problems summarized here constantly change. The areas offered for lease change. The number of auctions change. The bid adequacy rules change. The mininum bid changes. The regulations for removing platforms change. New fisheries and vacation destinations emerge. Yet the federal government, through one agency or another, has dealt with private sector use of the public domain for hundreds of years. The following section provides historical perspective on the process of land disposal and withdrawal.

HISTORY OF LAND DISPOSAL AND WITHDRAWAL

The federal process for transferring land to individuals and companies has existed since the nation began. This process has come to be called disposal. The source of lands for government disposal began onshore with the Articles of Confederation, which ceded western areas held by the original thirteen colonies to the new federal government. Since that time, the government has chosen to use its ownership of land in a variety of ways.

The first purpose of onshore land management by the government was to dispose of the property to private ownership. Varieties of disposal methods were employed, including outright sales, grants for building roads, canals, or railroads, for the services of military veterans, and for schools, among other purposes.[3] The second purpose of disposal was to raise revenues. Conflict began as a result of these two purposes early in this disposal process. The established states sought to reduce tax burdens by encouraging the government to sell the federal land, whereas the new territories and prospective settlers and merchants wanted free disposal of land. Rights to agricultural land were eventually sold in perpetuity (subject to tax payments) to private owners. In contrast, rights to mineral land, and in some cases the mineral holdings below private farm land, evolved into a leasing system. The significance of a leasing system is that the government does not relinquish claims forever but only for a period of time.

The evolution to a leasing system was a milestone separating free access and disposal to the private sector from continuous government management.[4] The Minerals Management Service represents the largest survivor of this historical disposal process. Its statutory basis, the Outer Continental Shelf Lands Act of 1953, requires that the offshore lands be leased.

At the turn of the nineteenth century, the states, particularly California, had begun to grant oil leases where onshore fields extended into shallow coastal waters. It was not until 1947 that a federal challenge to state leasing reached the U.S. Supreme Court. In an interesting example of judicial and congressional interaction, the Court found that the federal goverment had the right to lease all offshore lands. After extensive debate, Congress reacted by passing the Submerged Lands Act (1953) and the Outer Continental Shelf Lands Act of 1953. The latter restored state control to offshore lands extending three nautical miles from the shore except for the boundaries of Florida and Texas, which extended three nautical leagues or approximately nine miles. Remaining offshore lands, the OCS lands, were put under federal management.

As in the era of western land disposal, the details of government regulation affect the outcome. For instance, acreage limitations of the Homestead Act of 1862 had substantial impacts on the development of the American West. So, too, do the regulations on ocean disposal affect the current and future development of the ocean frontier. Among these limitations are the size of the claims on the Outer Continental Shelf, which are limited by law to be at most nine square miles, equivalent to 5,760 acres.

The conceptual similarity between western land disposal and current ocean leasing is visually striking. Consider Figure 1-1. The blocks represent land claims, the first in the Central Valley of California and the second off a portion of the coast of Texas and Louisiana. At the same time, important differences exist such as the distinction between first-come, first-claimed process in the West and the auction process by which offshore claims are recognized.

The philosophy of land management in the United States has evolved from one concerned totally with disposing federal land to one that includes reserving land for special uses. This evolution has required a changing view of nature. In particular, our country's view has changed from primarily viewing nature as an implacable and hostile environ-

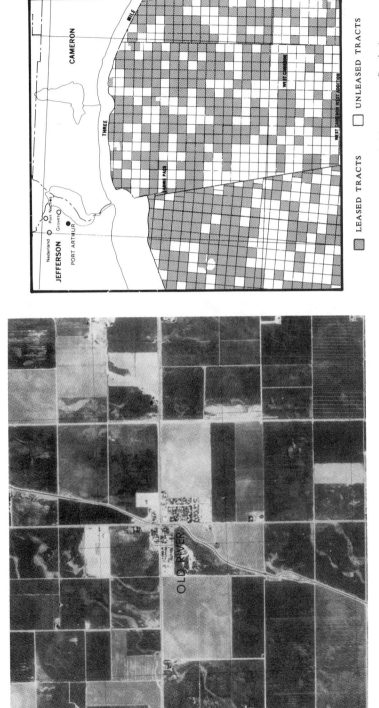

Figure 1-1. Land claims: Central California and the Gulf of Mexico. (*Sources*: U.S. Geological Survey; Minerals Management Service)

ment to be conquered to a popular view of nature as a necessary part of our cultural well-being.[5]

Minor withdrawals of federal land—land that could not be settled or claimed for private use—began in small ways by reserving some salt areas, a stand of trees for military ship building uses and hot springs such as what has become Hot Springs National Park in Arkansas. A more extensive withdrawal was the creation of Yellowstone Park (mostly in northwestern Wyoming) in 1872. Withdrawals reached substantial proportions beginning with the Forest Reserve Act of 1891, which led to the current system of national forests. Modern-day withdrawals are reflected in the creation of the wilderness system, scenic waterways, and wildlife sanctuaries. These withdrawals historically have created conflict between organized groups.

The withdrawal process is also ongoing and contentious in offshore lands. Some areas have been withdrawn permanently, such as those off Point Reyes in California, marine sanctuaries for the Channel Islands offshore Southern California, or the U.S. Monitor Sanctuary off North Carolina. There have been, however, very few of these permanent withdrawals. Instead, a compromise method has evolved that leads to temporary withdrawals called deferrals from leasing. These deferrals are a negotiated process at three different points in the management of the OCS.

The first point for negotiated withdrawals is at the development of the congressionally mandated five-year leasing plan prepared by the Department of the Interior. Historically, the second point, during the planning for a specific lease auction, has been a more important source of deferrals. At that point consultation with states, local governments, industry, and the public can lead to a deferral until the next auction in that area. The third point for negotiated deferrals has occurred during the annual appropriations process in Congress. From 1982 to the current time Congress attached an annual moratorium on leasing specific areas to the appropriations bill of the Department of the Interior. Political action after the 1989 spill from the *Exxon Valdez* led to increasing the area under the annual moratorium.

The historical process of disposal and withdrawal has taken place in a time of rapidly expanding human knowledge and institutions. Earlier eras could not even conceive of using petroleum products, much less extracting them from beneath the ocean, because of the limited state of human knowledge. Since the obverse of what is known is uncertain, it is the change in knowledge and the resulting reduction in uncertainty that substantially shapes the possibilities open for OCS management. The following section illustrates the importance of what is uncertain and summarizes the current state of knowledge in areas that are particularly critical to OCS management.

UNCERTAINTY: GEOLOGY, TECHNOLOGY, ECOLOGY, POLITICS, AND ECONOMICS

Management of the Outer Continental Shelf involves many dimensions of uncertainty, including uncertainties about the presence of petroleum, the effects of disturbing the ocean environment, changes in technologies for extraction and transportation, and the political and economic climates. The federal OCS managers are not the only ones facing these uncertainties. Managers in oil companies, state and local government managers, and individuals concerned about the environment face some or all of these dimensions of uncertainty.

Events affecting the California coastal area during the 15-year period from 1972 to 1987 can illustrate the reality of these uncertainties. Many, though not all, of the actions listed below are the delayed response and resolution of issues associated with the Santa Barbara oil spill of 1969. That spill, which occurred from Platform A located on a federal lease just over three miles offshore from Santa Barbara, helped usher in a nationwide era of environmentally focused legislation.[6]

The major events on the California OCS from 1972 to 1987:

1972—General price controls continue in effect from 1971. Coastal Zone Management Act becomes law.

1973—Middle East war and Arab oil embargo lead to the cost of imported oil rising by more than 4.5 times from 1972–1975.

1974—California State Lands Commission relaxes moratorium on drilling on state lands though no new sales are held.

1975—First federal lease sale since the 1969 oil spill in the Santa Barbara Channel. Strategic Petroleum Reserve established.

1976—Price freeze on petroleum products.

1977—The first development well into the Monterey Formation offshore is drilled into the Hondo field.

1978—Natural Gas Policy Act begins to decontrol the price of natural gas in interstate markets.

1979—Revolution in Iran leads to the cost of imported oil rising by more than 2.5 times in the 1979–1981 period.

1980—Crude Oil Windfall Profits Tax implemented. County legislation restricts passage from the Santa Ynez field.

1981—Crude oil and petroleum product prices are decontrolled. Point Arguello field discovered. Litigation on lease sale 53.

1982—Congressional moratorium on leasing in some areas alters lease sale 73.

1983—and 1984—Congressional moratorium continued.

1985—Congressional moratorium defeated. Rapid petroleum price decline begins and continues into 1986.

1986—The California congressional delegation and the Department of Interior seek an alternative to moratoria.

1987—Final five-year leasing plan passes with no lease sales in California until 1989.

Geologic uncertainty appears in two places in the above list. The first place is in 1977 when production first begins in the offshore extension of the Monterey rock formation. This formation is unusual in that it is both a source and a reservoir rock (discussed in the following section on geology). The second indication of uncertainty related to geology is that it was not until 1981 that the giant Point Arguello field offshore from Point Conception was discovered.

Technological uncertainty rarely shows up directly, though in 1979 the decision by county officials not to allow a pipeline from Platform Hondo off the California coast led to the use of an offshore storage and transfer ship, thus bypassing the need for a pipeline. Implicitly, however, technology changed throughout this time period as prospects in deeper water became technologically feasible and more wells could be drilled from a single platform among other changes during the period. Furthermore, technological choices, such as the decision to use a processing ship instead of a pipeline, often affect other uncertainties since the economics of alternative technologies and their risks differ.

Ecological uncertainty is as large as in the other areas. Major changes in the ecological landscape occurred during this period as the California brown pelican and the California sea otter expanded their range though other species of birds, reptiles, and mammals remain endangered. Ecological research begun during this period for the new requirements of Environmental Impact Statements continues with millions of dollars spent by the Minerals Management Service in the California area alone.

Political and economic uncertainty is rampant in the preceding list. Many of the listed items relate to political and regulatory changes resulting from changes in the law such as the political responses to the two energy crises beginning in 1973 and 1979, the organization of the California State Coastal Commission, and the annual battles over moratoria on leasing off California and other environmentally sensitive coasts. Finally, evidence for uncertain economics abounds as prices for petroleum increased and decreased dramatically due to decisions in the Middle East and political decisions regarding taxes and price controls.

These instances of uncertainty affecting management of the California Outer Continental Shelf are applicable more broadly to the management of the entire OCS. In order to understand the context of management decisions about the Outer Continental Shelf, it is necessary to understand the limits of the current knowledge of geology, technology, ecology, politics, and economics. These limits of current knowledge define the start of uncertainties for managing the Outer Continental Shelf. The following sections briefly survey the current boundary between knowledge and uncertainty in geology, technology, ecology, politics, and economics as applied to management of the Outer Continental Shelf.

Geology

Geological knowledge accepts changes in the earth that take place over hundreds of thousands, even millions of years. During that time mountains rise, oceans fall, and glaciers advance and retreat. Of particular importance to management of the OCS is the current understanding regarding sources and accumulations of hydrocarbons, the chemical family that includes petroleum and natural gas.

In general, current exploration and extraction of hydrocarbons is guided by geologic knowledge about sources of hydrocarbons (the source rock) and a geologic storage place for the hydrocarbons (the reservoir rock and trap). In both cases, geologists look to sedimentary rock created by the layering of sediments of all types over periods of millions of years.[7]

The source of hydrocarbons is thought to lie in what used to be major depressions that collected biological material (onshore or offshore) and sediments such that the biological material could not completely decompose. As an example, one could think of a deep estuary or coastal trough, low on oxygen, into which large amounts of biological matter and sediments accumulate over time. Throughout geological time, other layers accumulate, compacting the material below. The material formerly at the bottom of the estuary is likely to begin as a clay or mud, and under increasing pressure and temperature, become compacted to form shale. During this process the biological matter undergoes a transformation to become hydrocarbons while being slowly expelled from the compacting source rock.

Even if hydrocarbons are formed, where they end up is highly uncertain. The hydrocarbons leaving the source rock may simply find their way to the surface. Many oil and gas seeps are known throughout the world (including those off Santa Barbara), which account for approximately 8 percent of the oil spilled into the ocean each year.[8] Alternatively, hydrocarbons may migrate to a reservoir rock, a rock that is permeable and not undergoing compaction. (We know migration is possible because the extraction of hydrocarbons is the result of oil and gas migrating through rock to reach the well.) The migration stops in a reservoir rock, often a sandstone, when a geologic condition traps the migrating hydrocarbons in a geologic structure.

Figure 1-2 illustrates several of the geologic structures in which hydrocarbons are known to accumulate. In general, there must be a cap rock that is impermeable to the migration of hydrocarbons, for instance, a layer of now compacted clay that has become shale and a structure. Structures create localized accumulations of hydrocarbons as a result of faulting that breaks the migration path, as on the right-hand side of Figure 1-2, or perhaps as a result of a barrier created by an impermeable salt dome which has pierced the sedimentary rocks as on the left side of the figure.

Why is this such an uncertain process? One reason is the age of the rocks. Current hydrocarbon production is said to come from the "younger" sedimentary rocks formed within the last 50 million years. Yet hydrocarbons are found under the sands of the Middle East, in offshore waters and on land. The reason is that the geography of millions of years ago looked very little like that of today. As recently as 25,000 years ago, the seas may have been 400 feet lower than their current "sea level."[9] Over longer time horizons, the continents themselves have shifted. As a result, current knowledge of the generation and accumulation of hydrocarbons can guide exploration, but a great deal of uncertainty remains about finding new locations of hydrocarbons.

Geologic history determines the specific location of a hydrocarbon deposit, though that location has important implications for OCS management. The probability of oil spills reaching shore, the degree of political opposition, and the cost of the platform

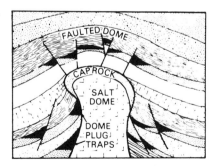 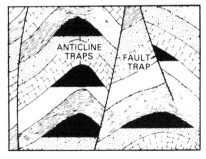

Figure 1-2. Geological structures. (*Source: Fundamentals of Petroleum*, M. Gerding ed., 3rd ed. 1986. Courtesy Petroleum Extension Service (PETEX), The University of Texas at Austin.)

and transportation system can all be affected by distance from shore among other factors. In the Gulf of Mexico, many production platforms are far out at sea, whereas off California most of the production platforms are easily visible from shore. The location of these platforms is driven by the geology.

The distance of the platforms from shore is affected by the width of the continental margin, the area including the Continental Shelf and the slope down to the deeper lands of the ocean, which varies greatly from coast to coast. The Continental Shelf is quite narrow off California, significantly wider off the Atlantic Coast and extends to a surprisingly great extent off Alaska. These differences are thought to result from the intersection of massive tectonic plates that make up the crust of the earth. In the Pacific, the Farrollon plate has been going beneath the North American continental plate. This has reduced the width of the shelf off that coast, which at places is virtually nonexistent, whereas in other places it may be as wide as 100 miles. The Atlantic Coast represents the trailing edge of the continental plate and extends farther under the ocean, whereas the continental margin in the Gulf of Mexico and Alaska extends many hundreds of miles into the sea.

Management decisions cannot shift the location of deposits, alter the width of the continental margin, or change other fundamental elements of the environment. But management decisions can use knowledge about the possible location of deposits and other environmental factors to shape and to constrain decisions.

Many aspects of the geologic processes remain so uncertain that the only way known to actually discover hydrocarbons is to drill for them. The following section surveys knowledge about the exploration and extraction of hydrocarbons that affects management of the OCS.

Technology

Existing technological knowledge constrains where it is desirable and possible to drill for oil and gas, how much can be extracted, the wastes that enter the environment from drilling and production, and the methods of transporting what is produced to shore. The technologies rely heavily on knowledge of electrical, mechanical, environmental, and civil engineering as applied to the ocean environment. Advances in any of these areas can alter the options that are open for management.

The extraction of oil, gas, or nonfuel minerals from the OCS requires the same sequence of exploration, extraction, and transportation. The technologies are similar for oil and gas, which are also the economically most important products. This section focuses on the technologies for oil and gas; Chapter 7 includes information on technologies for exploration and extraction of nonfuel minerals.

Exploration technologies are intended to provide information on the location of hydrocarbon deposits. These technologies cover a wide range of techniques ranging from the collection of seismic data (discussed below) to drilling. Federal regulations in general prohibit drilling until a lease has been obtained and permitting requirements satisfied. Prior to that point, as a guide both for exploration and for bidding for leases, companies may try to detect geologic traps through differences in magnetic and gravitational fields and through seismic surveys. The technology for seismic exploration, the most important of these methods prior to drilling, requires studying the pattern of waves propagated in the water, which are bounced off sedimentary layers at differing rates. The source of the waves may be electrical or acoustical. Significant advances have been made in the analysis of this information, which maps out, in somewhat fuzzy detail, the layers and major fractures of sediments.[10] Figure 1-3 indicates the patterns of waves as reconstructed by a computer for areas near the Baltimore Canyon off the coast of New Jersey. Also shown are the location of two wells drilled by Shell Oil Company, costing millions of dollars, which failed to locate commercial quantities of hydrocarbons. Current advances in computer technology are increasing the resolution of such profiles and reducing the labor-intensive interpretation of seismic data.

Drilling technology is fundamentally different and more expensive than seismic technology, although drilling may take place as part of exploration, development, or production. Rotary drilling technology has allowed wells to be drilled in depths of water as great as 7,000 feet and routinely drill several miles into the earth. Figure 1-4 represents several basic components of the circulation system involving a rotary device (the kelly), support for the drilling stem, and a hollow bit through which "mud" can be forced to lubricate and cool the bit while removing the drilling cuttings.

Many subsystems complicate this basic picture. The mud itself is part of a circulating system that removes the cuttings while allowing different additives to be included. The mud also plays the important role of maintaining pressure down the drill hole. One type of "blowout" associated with exploration occurs when the mud is not heavy enough to keep an oil and gas reservoir under control. If the pressure within the reservoir is too high, the mud can be forced out of the well followed by the oil and gas, a gusher in the old days, a management disaster today. The Minerals Management Service has also supported studies of the environmental effects of the muds and the cuttings that are brought up from the well.

In more detail, the exploration or production well is composed of a sequence of pipes within pipes. This ensures that hydrocarbons flow into the pipe and are captured by the operator instead of simply migrating to another layer of sediments, a possibility if the well is not sealed above the pay zone of the reservoir rock. The Santa Barbara oil spill was primarily the result of striking a high pressure reservoir prior to sufficient pipe

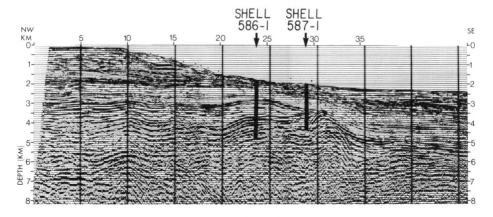

Figure 1-3. Seismic profile. (*Source:* U.S. Geological Survey, Woods Hole Group, courtesy of James Schlee.)

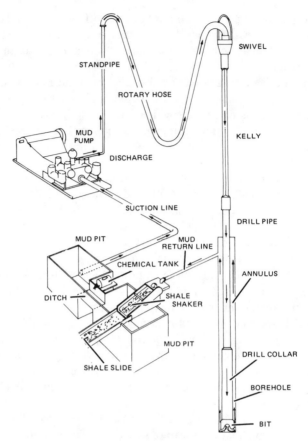

Figure 1-4. Circulation system. (*Source: Fundamentals of Petroleum,* M. Gerding, ed. 1986. Courtesy Petroleum Extension Service (PETEX), The University of Texas at Austin.)

being set to control the flow of hydrocarbons. As a result, much of the oil that was spilled came from oil that migrated from the reservoir rock 3,000 feet deep to shallower rock which then allowed passage to the ocean floor as a result of the extensive rock fracturing in that area.[11]

One accident control system is a blowout preventer illustrated in Figure 1-5. An extreme response to the loss of control of a well is to close the upper end of the pipe. The blowout preventer has several stages of closure ranging from sealing the outer pipe through which the drill stem travels to cutting off the drilling stem (pipe) itself and sealing the entire set of pipes.

The most visible aspect of exploration and production technology is the platform from which wells are drilled and maintained. Platforms used for the exploration phase are typically mobile. The drilling rig may be placed on a barge, a ship, a platform that sets down "legs" when in position (a jack-up rig), or a semisubmersible that may be self-propelled or require towing. Figure 1-6 illustrates each of these platforms as well as a fixed platform for production.

If exploration is successful, a permanent platform may be emplaced. Modern platforms may be the terminal point for over 60 individual wells that are angled away

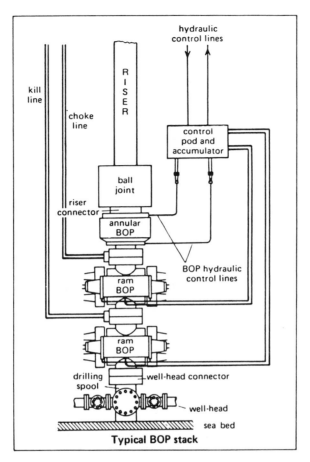

Figure 1-5. Blowout preventer (BOP). (*Source:* H. Whitehead, *An A–Z of Offshore Oil and Gas,* 2nd ed. Copyright © 1983 by Gulf Publishing Co., Houston, TX. Used with permission. All rights reserved.)

from the platform. This aggregation of wells reduces the need for separate platforms, thus decreasing costs and the number of visible structures. However, the increase in the number of wells on a single platform increases the risk of problems such as spills or fires in one well spreading to another well.

With the production of oil or gas comes a new set of technological issues that affect management. In particular, hydrocarbons must be monitored, transported, and processed, whereas occasional reworking of the well itself is necessary. The actual resource extracted from a well may contain substantial amounts of oil, gas, sand, and water intermixed. The various components must be separated either on the platform or at a shore-based facility. The unwanted material and water (produced water) is a waste product and may be dumped into the ocean after de-oiling or, more seldom, reinjected into the reservoir or barged to shore for land disposal.

Transportation on the OCS typically takes the form of pipeline transfers to a shore-based facility. More recent pipelines are buried, whereas older pipelines may lie on the surface of the seabed. Figure 1-7 is map of the pipelines connecting a few of the more than 2,500 platforms in the Gulf of Mexico. Whereas this transportation mode is out

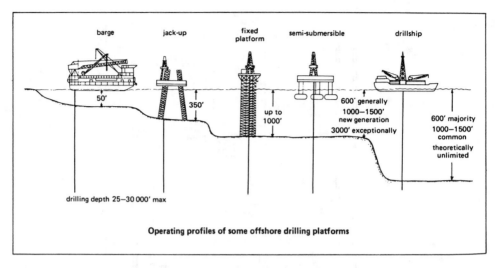

Figure 1-6. Drilling platforms. (*Source:* H. Whitehead, *An A–Z of Offshore Oil and Gas,* 2nd ed. Copyright © 1983 Gulf Publishing Company, Houston, TX. Used with permission. All rights reserved.)

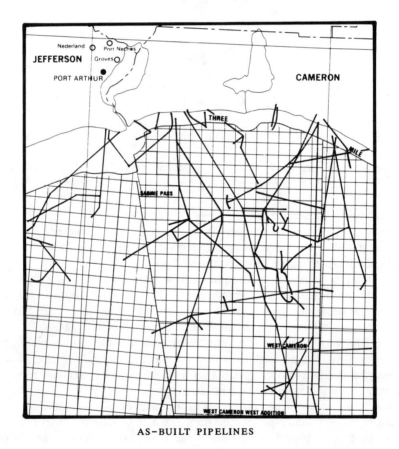

AS–BUILT PIPELINES

Figure 1-7. Pipelines near the Texas-Louisiana border. (*Source:* Minerals Management Service.)

of sight, it is not free from controversy as leaks may go unnoticed, fishing nets may tangle, or anchors may strike the pipeline.

Pipelines are, however, a safer method of transportation than by tanker ships, a topic discussed in more detail in Chapter 4. Tanker operations and accidents (part of transportation in Table 1-3) account for almost one-third of the total petroleum spilled into the ocean.

The mechanical, chemical, or electrical aspects of a technology are important to OCS management as they determine the costs of exploration, the probability of finding hydrocarbons, the amount that can be extracted, and the probability of spill. However, knowledge of technical details is not the only issue in the management of technology. Experience in situations ranging from airplanes, to nuclear reactors, to piloting ships has indicated that in crisis situations human reactions and decisions are an important determinant of the success or failure of the technology. A blowout preventer did exist on Platform A in Santa Barbara though it was not activated in time to stop the blowout of that well, nor is it clear that it could have been activated in time.

More recently, the 1989 spill in Alaskan waters from the *Exxon Valdez* illustrates further the uncertainties associated with the management of technologies. The *Valdez* was transporting oil that had been produced from Prudhoe Bay in Alaskan state waters and then transported through the Alaskan pipeline to the port of Valdez where it was loaded into the tanker. Contemporary accounts indicate a human error combined with tanker design may have been the immediate cause of the spill.[12]

The national anguish at the wreck of the *Exxon Valdez* seems prompted by fears of the effect of that massive, 11-million-gallon oil spill on the ecology of the area. Oil spills and their effects are discussed in more detail in Chapter 4; the following section discusses the current practice of ecology as it relates to Outer Continental Shelf management.

Ecology

Two pictures compete for the simplified view of the ecological effects of energy production from the Outer Continental Shelf. One is the picture many people carry in their minds of dead and oil-soaked sea birds. Another equally simplified picture shows sea lions sunning on a raft near an oil platform.

Table 1-3
Major Sources of Oil in the World Oceans: Best Estimates

	Million Metric Tons Annually	Percentage[a]
Natural sources	.25	8
Offshore production	.05	2
Transportation	1.47	46
Atmosphere	.30	9
Municipal and industrial waste and runoff	1.18	37
Total	3.2	

[a] Details do not add up to 100 due to rounding.
Source: National Research Council, *Oil in the Sea: Inputs, Fates, and Effects* (Washington, DC: National Academy Press, 1985), p. 82.

The actual effect of OCS operations on ocean and coastal ecology is more complex and subject to more uncertainty than the simplified view suggests. Research on the ecological effects of oil and gas operations is region specific and in the past has focused on the effects of discrete events such as oil spills. Such spills seem to have been the energizing cause of activism by many people, and so are extensively studied.

Not only have specific spills been studied, but the Minerals Management Service has maintained an ongoing environmental studies program in support of its operations. Research from these studies has been used in the environmental impact statements that are necessary for five-year plans as well as for individual sales.

The environmental impact studies are region specific and are intended more for an individual auction than for a five-year plan. Each impact statement generally surveys the existing benthic (bottom-dwelling) plant and animal community, the species in the water column, the species that migrate through the area, and those that use both the shore and the sea. Substantial attention is devoted to species of particular concern such as sea otters and whales. For the species studied, there are projected impacts on the populations from oil spills of various magnitudes.

Formal modeling of the effects of oil spills involves studying a source of the spill, the transport of the spill through the various parts of the water column and into the atmosphere, the fate of the spill both in terms of its chemical transformation and its interaction with living elements of the environment, and, finally, its effects. For policy purposes, effects are often those valued in the economy even though many intermediate effects have been studied.

In general, an oil spill has an immediate impact on the surrounding ecology. Sea birds and sea otters do die from loss of flotation and warmth. Shellfish concentrate hydrocarbons and can become unsafe to eat. Plankton, the building block of the ocean food chain, die with commensurate effects throughout the chain. The bulk of research indicates, however, that damage to an area from an oil spill depends on the severity of the oiling, and recovery from severe oiling occurs even though substantial recovery may require 5–10 years.[13]

In contrast to the study of discrete events such as oil spills, less work has been done on the persistent effects of low levels of oil in the sea, although many biological processes are known to be sensitive to low levels of petroleum. Particularly in the Gulf of Mexico and increasingly in California, it is possible that persistent low levels of pollution permanently alter the ecology. The Minerals Management Service has recently entered into contracts, one with the Louisiana Universities Marine Consortium and the other with the University of California at Santa Barbara to maintain long-term studies on the possible persistent effects of low level pollution.

Various aspects of ongoing oil and gas operations have been studied for their effect on the ocean and coastal ecology. Topics include: the effects of seismic exploration on fishing populations, the effect of drilling muds and cuttings, the loss of wetlands due to onshore development associated with OCS operations, and the effect of the platforms themselves.[14] Platforms alter at least the location and perhaps the quantity of fish in the Gulf of Mexico by providing a complex structure in the otherwise undifferentiated shelf. In contrast, the presence of platforms off California does not seem to affect fish concentrations as much on the short and rocky shelf off California. In Alaska, a major concern was that the platforms would interfere with the migration paths of whales.

The evidence to date for these nonspill effects on the environment indicates that the effect is either very local, as for the presence of drilling muds and cuttings, or temporary. The possibility of longer term effects exists for the loss of wetlands, particularly along the Gulf of Mexico.

Finally, there are concerns about the effect on air quality of emissions of oxides of nitrogen and sulphur and reactive hydrocarbons from operating platforms. These chemicals are important contributors to the formation of ozone and are of particular concern off Southern California where the nearby Los Angeles basin is a nonattainment area for ozone under regulations stemming from the Clean Air Act. Atmospheric discharges on the OCS are currently under the regulatory authority of the Minerals Management Service, which has instituted regulations intended to avoid significant onshore air quality impacts of offshore operations. However, ongoing activity exists to either transfer this regulatory control to the Environmental Protection Agency, which administers other nationwide regulations for atmospheric emissions, or to equate onshore and offshore regulations.[15]

Whereas offshore operations do affect the environment, the effect has not proven to be so devastating that all offshore operations are halted on the basis of scientific evidence. Uncertainties and risks do remain to the ocean ecology from offshore drilling, however, and individual perceptions of damage to the environment may vary greatly. Politics therefore plays an important part in management of the OCS when ecological beliefs clash with the economic effects of offshore oil and gas leasing.

Politics

Participation in the political process is often the most direct expression of feeling for individuals affected by activity on the OCS. As a result, political sensitivity of OCS management appears in several ways. One way is through the organizational structure of the government. Organizationally, the Minerals Management Service is a part of the executive branch of government. As such, its senior policy officials are political appointees who report to the politically appointed assistant secretary for energy and minerals in the Department of the Interior.

Another indication of the political sensitivity of OCS management is the interaction among federal agencies. Particularly important interactions are among the Minerals Management Service, the Environmental Protection Agency, and the National Oceanic and Atmospheric Administration, a part of the Department of Commerce. The previous section on ecology introduced the idea of agency conflict over the release of atmospheric pollutants. The National Oceanic and Atmospheric Administration can find itself in conflict with management practice on the OCS through its role as the overseer of the coastal zone management programs of the states. A further conflict exists over a state's management of its coastal zone and federally managed activities on the Outer Continental Shelf.

The conflict between coastal zone management programs and federal management of the OCS has recently revolved around the sharing of OCS revenues and the so-called consistency doctrine regarding energy production and coastal zone management plans. It is not surprising that coastal states want a larger share of the funds that are earned off their coasts but go into the federal treasury. The states recently received the right to 27 percent of the funds received by the government in lands—the so-called 8(g) lands—three additional miles beyond their coastal waters. More contentious is the requirement that actions on the OCS be consistent with federally approved, coastal zone management plans. Important litigation, discussed in Chapter 2, exempted the five-year plan and auctions from being consistent with coastal plans. This naturally sets up political conflict between various agencies and political actors.

A second way in which the political sensitivity of OCS management appears is through elected officials and the election process. In the national political debates of 1988, citizens in several coastal areas such as northern California and North Carolina were exhorted to "vote the coast." This was interpreted to imply a vote against any offshore development. Strong political feelings such as these translate into congressional action. Recent examples are the moratoria on offshore leasing off California and Massachusetts passed as part of annual budget legislation in the years since 1982.

Political views are, therefore, important to current practice within the Minerals Management Service, among federal agencies, in regional politics, and in national politics. Some, not all, of the political issues can be resolved on the basis of improved information. It is clear, however, that some irreducible amount of conflict over alternative uses of the Outer Continental Shelf will remain.

Economics is the study of the allocation of scarce resources among alternative uses. Though resources and uses are considered to have very general definitions in economics, management of the OCS clearly requires difficult choices that allocate the resources—energy, recreation, water, fisheries—among alternative uses.

Economics

Economic issues affect the current practice of OCS management in a variety of ways. Fundamentally, leasing occurs only because companies believe it is profitable to extract oil and gas. This belief naturally changes as the price of oil and gas changes and as the costs of exploration, production, and transportation change. The economics of oil spills—to the company doing the spilling and to the other companies, states, and individuals who are damaged—affects OCS management. Property values, tourist activity, and income from fishing are all economic issues that affect OCS management.

For companies that seek to extract oil and gas from the Outer Continental Shelf, the current economics of offshore energy production requires that oil and gas deposits exceed a minimum size and be in sufficiently shallow water to be profitable. The minimum size and water depth vary by region as well as changing over time as technologies and prices change. Listed below is the range of minimum economic field sizes, in million barrels of oil, used in a recent study of OCS resources.[16]

Gulf of Mexico	3–690
Pacific	3–190
Alaska	45–300
Atlantic	5–1,000

As the prices of oil and gas rise and fall, so does interest in the resources of the Outer Continental Shelf. As indicated in Table 1-2, a doubling of price leads to a fourfold increase in the quantity of resources that could be profitably extracted from the OCS.

In addition to the private economics of profit from oil and gas extraction, the Minerals Management Service considers, for some analyses, additional social costs of offshore energy production. These social costs are those incurred by other members of society who are not a part of the direct purchase and sale of hydrocarbons. In economics, these additional social costs are called external costs. The costs are external to the companies extracting the resources, but economists argue that they are part of the total cost of production. These costs primarily include the expected value of damages from oil spills (discussed in Chapter 4) but could also include changes in the value of onshore property caused by oil and gas operations and other activities such as recreation, which could also be affected.

Fundamentally, however, economics provides a structure for the analyses of management policies. Economics provides one way to aggregate some effects of geology, technology, ecology, and individual preferences in order to evaluate alternative policies. At the same time, uncertainty about the economic variables such as future prices and costs provides a challenge to managers of the OCS.

Chapter 2

Institutions and Goals

Major institutions define the goals of Outer Continental Shelf management and modify its management practices. Whereas the Minerals Management Service has been described as the government institution with actual management responsibility, many other institutions affect its management. Congress created the statutory and budgetary authority for the Minerals Management Service. In that statutory authority, Congress has given broad powers to the secretary of the Interior, but also established processes for participation by states, local governments, interest groups, and other institutions. This chapter describes the statutory goals and the key institutions in the management of the Outer Continental Shelf.

Managing is difficult without a clear purpose for management. For private companies, pursuit of profit is the major purpose; the purposes of governmental management are more obscure due to conflicting objectives and difficulties of measurement.[1] In practice, the management of the OCS reflects purposes from three sources: the goals specified by Congress, the organizational goals of the Minerals Management Service, and the goals of institutions in the broader ocean management system. Each is considered in the following sections, concluding with a discussion of how policy analysis attempts to unify practice and purpose.

CONGRESSIONAL GOALS AND THEIR INTERPRETATION

Legal authority for executive branch agencies such as the Minerals Management Service derives from congressional action. The history of congressional action that led to the major enabling statutes for OCS management[2] is described in Chapter 1. These statutes are typical of the broad delegation of authority by Congress to an executive branch agency. The acts list broad and often conflicting goals, such as expediting exploration and development and minimizing the risk to the marine, human, and coastal environment. The statutes provide the more detailed directions on auctions outlined in Chapter 1, though many specifics of implementation are delegated to the secretary of the Interior. As a result, the interpretation of the intent of Congress is an ongoing management problem in the Minerals Management Service, an issue in judicial decisions, and a topic for congressional amendments and investigation.

To further complicate management, many other laws that specify goals must also be considered. The Minerals Management Service has published a list of 76 other laws that affect OCS management, including the Clean Air and Clean Water acts, the Coastal Zone Management Act, the Endangered Species Act, the Mineral Leasing Act of 1920, and the Natural Gas Policy Act of 1978 among others.[3] This section focuses, however, on the statements of congressional intent in the OCS acts and the degree of scrutiny that the judiciary applies to administrative interpretations of this intent.[4]

The broadest congressional statement of OCS policy is contained in the Outer Continental Shelf Lands Act (OCSLA) where it is stated:

It is hereby declared to be the policy of the United States that—

(3) the Outer Continental Shelf is a vital national resource reserve held by the Federal Government for the republic, which should be made available for expeditious and orderly development, subject to environmental safeguards, in a manner which is consistent with the maintenance of competition and other national needs;

(4) since exploration, development, and production of the minerals of the Outer Continental Shelf will have significant impacts on coastal and non-coastal areas of the coastal States, and on other affected States, and, in recognition of the national interest in the effective management of the marine, coastal, and human environments—

(A) such states and their affected local governments may require assistance in protecting their coastal zones and other affected areas from any temporary or permanent adverse effects of such impacts; and

(B) such states, and through such states, affected local governments, are entitled to an opportunity to participate, to the extent consistent with the national interest, in the policy and planning decisions made by the Federal Government relating to exploration for, and development and production of, minerals of the Outer Continental Shelf;

(5) the rights and responsibilities of all States and, where appropriate, local governments, to preserve and protect their marine, human, and coastal environments through such means as regulation of land, air and water uses, of safety, and of related development and activity should be considered and recognized; and

(6) operations in the Outer Continental Shelf should be conducted in a safe manner by well-trained personnel using technology, precautions, and techniques sufficient to prevent or minimize the likelihood of blowouts, loss of well control, fires, spillage, physical obstruction to other users of the waters or subsoil and seabed, or other occurrences which may cause damage to the environment or to property, or endanger life or health. 43 USC 1332

These broad goals, as well as more detailed text on some specific issues, do provide grounds for discussing the trade-offs involved in any given policy (though how one trades off between the categories is not specified). However, Congress provided a more specific declaration of purpose in the 1978 amendments, stating that the purposes of the act were to:

(1) establish policies and procedures for managing the oil and natural gas resources of the Outer Continental Shelf which are intended to result in expedited exploration and development of the Outer Continental Shelf in order to achieve national economic and energy policy goals, assure national security, reduce dependence on foreign sources, and maintain a favorable balance of payments in world trade;

(2) preserve, protect, and develop oil and natural gas resources in the Outer Continental Shelf in a manner which is consistent with the need

(A) to make such resources available to meet the Nation's energy needs as rapidly as possible,

(B) to balance orderly energy resource development with protection of the human, marine, and coastal environments,

(C) to insure the public a fair and equitable return on the resources of the Outer Continental Shelf, and

(D) to preserve and maintain free enterprise competition;

(3) encourage development of new and improved technology for energy resource production which will eliminate or minimize risk of damage to the human, marine, and coastal environments;

(4) provide States, and through States, local governments, which are impacted by Outer Continental Shelf oil and gas exploration, development, and production with comprehensive assistance in order to anticipate and plan for such impact, and thereby to assure adequate protection of the human environment;

(5) assure that States, and through the States, local governments, have timely access to information regarding activities on the Outer Continental Shelf, and opportunity to review and comment on decisions relating to such activities, in order to anticipate, ameliorate, and plan for the impacts of such activities;

(6) assure that States, and through States, local governments, which are directly affected by exploration, development, and production of oil and natural gas are provided an opportunity to participate in policy and planning decisions relating to management of the resources of the Outer Continental Shelf;

(7) minimize or eliminate conflicts between the exploration, development, and production of oil and natural gas, and the recovery of other resources such as fish and shellfish;

(8) establish an oil spill liability fund to pay for the prompt removal of any oil spilled or discharged as a result of activities on the Outer Continental Shelf and for any damages to public or private interests caused by such spills or discharges;

(9) insure that the extent of oil and natural gas resources of the Outer Continental Shelf is assessed at the earliest practicable time; and

(10) establish a fisherman's contingency fund to pay for damages to commercial fishing vessels and gear due to Outer Continental Shelf activities.

These congressionally mandated purposes seldom specify the exact program or the desired trade-offs to achieve their ends, or even the quantifiable measures to be used to evaluate the trade-offs. It is this lack of specificity that requires administrative and judicial interpretation. Congress delegated the role of administrative interpretation and power to the secretary of the Interior who in turn delegated day-to-day management to the Minerals Management Service.

Members of the Minerals Management Service are constantly engaged in interpreting the congressional intent of these broadly stated goals. Affected members of the public, however, can take the Minerals Management Service and the Department of the Interior to court if they believe that they are failing to follow the statute, however vaguely defined. This leads to federal judicial review—federal since the United States is a party—of agency interpretations of statutes. In general, substantial precedent for judicial review of agency programs as well as a statutory basis in the Administrative Procedure Act of 1966 exists. That act states that agency actions may be set aside by the courts if the actions are "arbitrary, capricious . . . (or) unsupported by substantial evidence."[5] Various standards of review can be applied to different aspects of a case; one standard being applied to the facts, a second to policy choices, and a third to statutory interpretation. Regarding the latter, in a recent important case concerning the OCS,[6] a federal appeals court reaffirmed that:

> (W)e adhere to the principle that the interpretation of a statute by those entrusted with its administration is entitled to substantial deference . . . (however) (a)n administrative interpretation of a statute which does not effectuate the intent of Congress must fall."

The Minerals Management Service and the Department of the Interior can therefore exercise substantial but not unlimited discretion in their interpretation of broad congressional goals. This leeway to interpret congressional goals allows an agency to achieve goals that some can interpret as being other than those specified by Congress.

THE MINERALS MANAGEMENT SERVICE

A large literature exists on the goals of organization managers. A frequently stated goal of governmental managers—to increase their budgets—may reflect managers' desires to manage large and growing organizations.[7] In the case of the Minerals Management Service, its financial operations are also of interest to other government organizations. Lease sales generate billions of dollars that are viewed as an important indicator of performance by the Office of Management and Budget. Alternative goals for an agency can also develop through an organizational ethic. A particular organizational ethic may develop over time that attracts certain kinds of managers and constrains to some degree the options open to management.[8]

The history of the Minerals Management Service is an important clue to its organizational goals. Its history begins with the Conservation Division of the U.S. Geological Survey, which supervised OCS exploration and production and conducted resource and economic evaluations from 1953 to 1982. The U.S. Geological Survey is a scientifically oriented organization,[9] and the Conservation Division, which held management responsibilities, formed the basis for the Minerals Management Service, created by a secretarial order in January 1982. In May of 1982, the supervision of the lease auctions and collection of revenues was also transferred from the Bureau of Land Management to the Minerals Management Service. Members of the OCS coordinating office at the department level were also transferred to the new agency.[10]

The birth of the Minerals Management Service coincided with the emphasis of the Reagan administration on market forces, and so policies changed to expand industry's choice of areas to lease and to estimate the value of fewer tracts. This reorganization and altered emphasis, however, was carried out with the same personnel originally transferred from the Conservation Division and the Bureau of Land Management. As a result, the Minerals Management Service has something of a split organizational ethic.

The carryover of personnel from the original agencies represents a distrust of market processes—leasing should be carefully controlled by those with specialized knowledge in government. The new organizational ethic from the early 1980s was to let the market provide the expertise and that, in effect, the OCS leasing program was a disposal program whose success might be measured by the extent of transfers to the private sector.

As seen in the organization chart of the Department of the Interior (Fig. 2-1), the Minerals Management Service is under the supervision of the assistant secretary for land and minerals along with the Bureau of Land Management, which administers the majority of the onshore federal domain and the Office of Surface Mining. Funding for the Minerals Management Service represents 2 percent of the department budget but is by far the largest source of revenue within the department and the largest source of income for the federal government outside the Treasury department.

The organization chart of the Minerals Management Service itself reflects some of its split historical personality. As detailed in Figure 2-2, the Minerals Management Service has four basic programs that may be in either the headquarters office or in field offices. These four programs are prelease activities (resource evaluation, including economics), postlease activities (operations), environment, and royalty management.

Table 2-1 presents the major expenditure categories, the budget authority, and the number of full-time personnel in the Minerals Management Service in 1987. These financial descriptions of the organization are discussed in more detail in Chapter 6.

GENERAL ORGANIZATION – U.S. DEPARTMENT OF THE INTERIOR

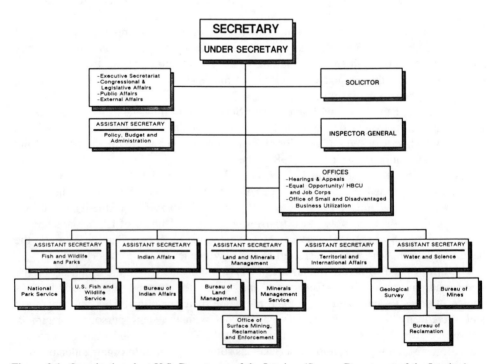

Figure 2-1. Organization chart U.S. Department of the Interior. (*Source:* Department of the Interior.)

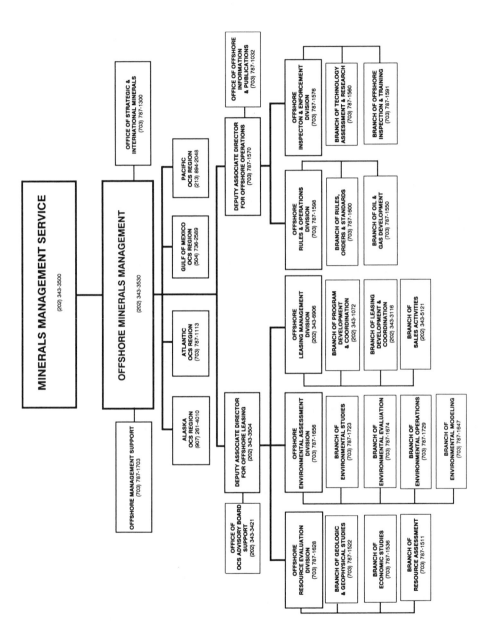

Figure 2-2. Organization chart Minerals Management Service. (*Source:* Minerals Management Service.)

Table 2-1
1987 Budget and Personnel, MMS

Major Expenditure Categories	Funds (millions)		Personnel	
OCS OPERATIONS				
Leasing and environment		38.6		343
Resource evaluation		26.2		340
Regulatory program		27.9		424
Subtotal	92.7			1,107
JOINT OPERATIONS[a]				
Royalty management		45.7		620
General administration		23.1		292
Subtotal	68.8		912	
Total	161.5		2,019	

[a] These expenditures and personnel refer to both OCS operations and the onshore royalty collection responsibilities of the Minerals Management Service. Onshore royalty collections comprise about 20 percent of the total collections of MMS. (*Source*: MMS Budget Justifications.)

STATE AND LOCAL GOVERNMENT AND INTEREST GROUPS

The theory of interest group politics focuses on the personal and organizational objectives of those for whom it is valuable enough to take part.[11] Each state, county, city, and interest group has somewhat different interests in affecting the management of the Outer Continental Shelf as well as different statements of purpose.

Historically, major players in this group taking active parts have included the states of California, Massachusetts, Texas, and Florida, and some cities, such as Santa Barbara. Many interest groups, including the Natural Resources Defense Council, the Sierra Club, Greenpeace, and the American Petroleum Institute among many others, have also actively participated in affecting the politics of OCS management. Because of the diversity of purpose among these interested groups, this section focuses on the ways in which the groups can and have taken part in the process of OCS management.

Advisory committees to the Minerals Management Service are one avenue of influence open to the coastal states and interest groups. Figure 2-3 is the organization chart for the OCS advisory committees. These committees include a policy committee, regional technical working groups, and a scientific committee. As the names indicate, there is an increasing specificity to each of the committees. The first two committees include members from each of the coastal states as well as appointments from other areas. Appointment to the scientific committee is based on other criteria including professional accomplishment.

The advisory committees are regularly briefed by and communicate with the Minerals Management Service. However, a number of other opportunities exist for states, local government, and interest groups to participate, with varying degrees of effect, in the management process.

Preleasing: State and Interest Group Opportunities

A substantial amount of interest group politics was built into the planning stages of the leasing process by the 1978 amendments. The original OCS Land Act in 1953 estab-

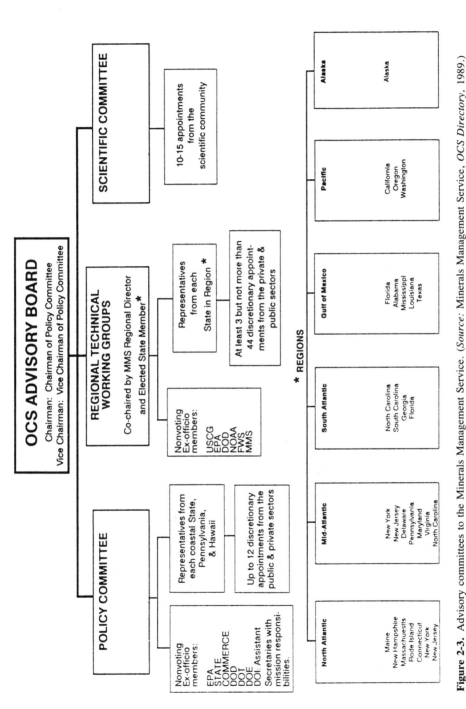

Figure 2-3. Advisory committees to the Minerals Management Service. (*Source:* Minerals Management Service, *OCS Directory,* 1989.)

lished that the OCS was an area of exclusive federal jurisdiction and determined that
state civil and criminal laws would apply and that OCS lands would be exempt from
state taxation.[12] In contrast, the 1978 amendments directed a substantially increased
advisory role for state and interest groups, and the newly required five-year leasing plan
also provided a mechanism for increased congressional, state, and interest group participation.[13] The five-year plan was specified in order to address a variety of goals
(many of which have already been discussed as mandated goals for the program in
general).[14]

Procedurally, however, the requirement to submit a five-year plan that addresses old
as well as new goals has substantially altered the planning process. Specifically, Congress required that a five-year plan be prepared by stating that:

> The leasing program shall consist of a schedule of proposed lease sales indicating, as
> precisely as possible, the size, timing, and location of leasing activity which he (the Sec-
> retary) determines will best meet national energy needs for the five-year period following
> its approval or reapproval. Such leasing program shall be prepared and maintained in a
> manner consistent with the following principles:
>
> (1) Management of the Outer Continental Shelf shall be conducted in a manner which
> considers economic, social, and environmental values of the renewable and nonrenewable
> resources contained in the Outer Continental Shelf, and the potential impact of oil and gas
> exploration on other resource values of the Outer Continental Shelf and the marine, coastal,
> and human environments.
>
> (2) Timing and location of exploration, development, and production of oil and gas
> among the oil- and gas-bearing physiographic regions of the Outer Continental Shelf shall
> be based on a consideration of—
>
> (A) existing information concerning the geographical, geological, and ecological
> characteristics of such regions;
>
> (B) an equitable sharing of developmental benefits and environmental risks among
> the various regions;
>
> (C) the location of such regions with respect to, and the relative needs of, regional
> and national energy markets;
>
> (D) the location of such regions with respect to other uses of the sea and seabed,
> including fisheries, navigation, existing or proposed sealanes, potential sites of deep-
> water ports, and other anticipated uses of the resources and space of the Outer Conti-
> nental Shelf.
>
> (E) the interest of potential oil and gas producers in the development of oil and gas
> resources as indicated by exploration or nomination;
>
> (F) laws, goals, and policies of affected States which have been specifically identified
> by the Governors of such states as relevant matters for the Secretary's consideration;
>
> (G) the relative environmental sensitivity and marine productivity of different areas
> of the Outer Continental Shelf; and
>
> (H) relevant environmental and predictive information for different areas of the Outer
> Continental Shelf.
>
> (3) The Secretary shall select the timing and location of leasing, to the maximum extent
> practicable, so as to obtain a proper balance between the potential for environmental dam-
> age, the potential for the discovery of oil and gas, and the potential for adverse impact
> on the coastal zone.

(4) Leasing activities shall be conducted to assure receipt of fair market value for the lands leased and the rights conveyed by the Federal Government. 43 U.S.C. 1344

Extensive litigation has been based on this portion of the Outer Continental Shelf Lands Act, particularly the California v. Watt cases discussed in Chapter 6. This litigation results in the current five-year plan specifically addressing each of these items. The preleasing process now consists of two phases: the five-year plan for an entire series of lease sales, and the detailed planning for an individual lease sale. Tables 2-2 and 2-3 highlight the opportunities not only for states and local governments but also for other interest groups to participate formally, though in an advisory capacity, in these planning processes. The first public notice for the 1992 five year plan appeared in 1989.

In fact, interested parties participated in the development of the 1987–1992 plan in a variety of other formal ways. Public hearings were held on the plan in several of the coastal states, a special congressional committee investigated alternatives for lease sales off California to which interested parties could direct comments, and, finally, Congress had a final review period in 1987 to respond as they saw fit.

Some of the detailed issues and negotiations behind the five-year plan are discussed in the next section. It is useful, however, to present the further planning and consultation steps for any one of the 36 lease sales scheduled in the five-year plan. Table 2-3 outlines the typical schedule for a standard lease sale.

Managerially, states and interest groups can formally and informally (depending on specific circumstances) advise the preleasing process.[15] One important question is the degree to which the secretary of the Interior and the Minerals Management Service must respond to the input that is provided. Again, there are formal provisions and informal realities.

For both a five-year plan and specific lease sales, the secretary must reply in writing to recommendations of the governors of the affected states. For the five-year plan, this correspondence is forwarded to Congress along with the final plan. For specific lease sales, the secretary is required to consider the balance between national interest and the well-being of the citizens of the affected state. However, the secretary's decision is considered final in the case of individual lease sales.

In regard to public comments received at the specified stages, the Minerals Management Service must summarize the issues raised but is not required to reply to each comment. A recent development, however, is the importance of participation in the planning process outside of the formal comment process, which is discussed below.

Table 2-2
Public Notices for the 1987 Five-Year Plan

July 11, 1984	5-year OCS Oil and Gas Leasing Program; Development and Request for Comment.
March 22, 1985	Request for comments on the Draft 5-year plan.
August 19, 1985	Request for comments on Appendix P: Analysis of Tract Selection and Areawide Leasing Approaches (an appendix to the 5-year plan).
February 7, 1986	Request for comments on the proposed 5-year plan.
April 25, 1987	Final proposed 5-year plan submitted to Congress.

Source: Department of the Interior, *Five-Year Leasing Program,* Appendix M.

Table 2-3
Participation Points for Standard Sales

Purpose	Month	Contact
Early coordination		Contact Governors, contact state members of the OCS Advisory Board Policy Committee, & the Regional Technical Working Group
Call for information and nomination, notice of intent to prepare EIS	1	Bidders to nominate areas of interest, States and interested parties to identify environmental effects and use conflicts. Provide information to states on shared, 8(g), land
Draft Environmental Impact Statement	12	Comments requested and hearings held for a 60 day period
Final Environmental Impact Statement	18	Report submitted to EPA and available to the public
Proposed Notice of Sale	19	Specifies terms and conditions, to States for 60 day comment
(remaining items do not allow comment but are public notices and key events)		
Final Notice of Sale	22	At least 30 days prior to sale
Lease sale	23	Public opening and reading of bids
Leases issued	25	After bid adequacy review and anti-trust review, leases issued within 90 days.

Source: Appendix L, *Five-Year Leasing Program* and *Oil and Gas Technologies for the Arctic and Deepwater,* Congress of the U.S., Office of Technology Assessment, 1985.

Preleasing: Deferrals

Withdrawing or deferring areas from leasing has become a major battleground between interest groups and the Minerals Management Service. Beyond state waters, the closest action to permanent withdrawals are the establishment of marine sanctuaries and the few withdrawals based on amendments to the OCS Lands Act. These latter withdrawals include areas for national defense use as well as scenic areas near Point Reyes seashore in California. The marine sanctuaries, once established, can be open to leasing only if the secretary of Commerce, not the secretary of the Interior, determines that leasing in a specific area is consistent with the protection and restoration of a sanctuary's conservation, recreation, ecological, or aesthetic value.[16]

More common, however, is altering the area offered for a particular sale after interest groups have commented on the initial area suggested by the Minerals Management Service. These alterations are called deletions or deferrals and are for a particular sale only. Five-year deferrals from leasing appeared for the first time in the 1987 five-year plan. (The plans in 1980 and in 1982 did not include deferrals.) The current five-year plan contains deferrals that are associated with military zones, buffer zones around selected areas such as marine sanctuaries and areas of special biological significance, major fisheries, particularly scenic or environmentally sensitive regions, and locations that are technically unfeasible to develop given the current price and cost conditions.

The deferrals included in the 1987 five-year plan were largely determined by political interventions utilized by states and interest groups outside of the standard commenting process. The interventions took the forms of congressionally mandated moratoria on leasing in a specified area, committee negotiations between Congress and the secretary of the Interior, and interest group negotiations promoted by the Institute for Resource Management.

Congressional moratoria, usually based on single year funding limitations specified by the appropriations committee, are the longest standing form of intervention. Congress first passed a moratorium on leasing selected areas on the OCS in 1982; some part of the OCS has been under a leasing moratorium every year since then. In 1982, the secretary of the Interior, James Watt, was proposing to significantly increase the area available for lease. The political reaction by members of Congress was so strong that substantial areas off the California and Massachusetts coasts were legally exempted from leasing for a period of one year. The geographic focus of these moratoria areas were the Georges Bank fishing ground off Massachusetts and the area off the coast of the scenic Big Sur peninsula in California. In all, over 70,000 square miles were exempted from leasing off the California and Massachusetts coasts. Over the years various parts of OCS lands have been added or subtracted from these moratoria. Most recently, following the spill from the *Exxon Valdez*, a moratorium on leasing was enacted that included areas from Maryland to Maine on the East Coast, all of California, parts of Florida, and, for the first time, part of Alaska. However, these moratoria and the substantial litigation to which the first two five-year plans were subjected directly affected specific lease sales and led to a desire by the Minerals Management Service to reduce the political conflict associated with its program.

Following the 1985 moratorium, the secretary of the Interior was instructed by Congress to consult with congressional representatives to determine if a compromise could be reached. This consultation led to a second major form of political intervention. Initially, in 1985 the secretary of the Interior, Donald Hodel, met with members of the California congressional delegation to work out a compromise between moratoria that deferred broad leasing areas and the desire of the Minerals Management Service to develop an ongoing leasing program. Though a preliminary agreement was reached that could have resulted in approximately 200 tracts off central and northern California being available for lease, the Department of the Interior withdrew from final commitment when a more detailed analysis indicated the relatively poor resource potential of the specified tracts.[17]

The failure of this ad hoc negotiating committee led Congress to select a committee to develop leasing alternatives for the California Outer Continental Shelf. The several alternatives proposed by this committee were incorporated into the five-year plan as alternative deferral options for consideration by the secretary. The final five-year plan delayed leasing in California until 1989, which was further delayed when President Bush appointed a top level, multiagency task force to study leasing off California and Florida and to report to Congress in 1990.

Though neither the moratoria nor the ad hoc committees is part of ongoing congressional participation in lease sale planning, these congressional interventions do indicate disagreement with the Interior's implementation of congressionally mandated goals.

Congress, particularly within the appropriations committees, is indicating that too little weight had been attached in specific areas to fishing, tourism, and environmental uses of the OCS.

Interest group negotiations begun by the Institute for Resource Management is a third example of deferral intervention. The Institute, a nonprofit group begun by Robert Redford in 1981 to develop methods to balance environmental and development issues, focused on the tentative plan for leasing in the Bering Sea off Alaska. This area contains the largest commercial fishery in the United States, with substantial environmental interest as well. The Institute for Resource Management convened a negotiation session of interested parties, including fishing associations, local governments, native associations, environmental groups, and oil and gas companies. This group presented its recommendations to Secretary Hodel in 1986. The final five-year plan deferred some of the areas specified by the Institute for Resource Management and highlighted others for further study. The work of the negotiating group was explicitly mentioned as a part of the planning process.

The signal from these interventions for management of the OCS is that collective action outside of the standard process can have an impact on decisions; formal economic or scientific analysis of the trade-offs between congressionally mandated goals is not necessary to have an effect on the timing and location of lease sales. It remains to be seen whether these strategies of congressional intervention and interest group coalition will be applied to deferrals for specific sales. These strategies are relatively new compared to public review and comment at specified points and compared with litigation when environmental impact statements and lease sales are finalized.

These politically focused methods of achieving deferrals all occur at the prelease, planning stage. Whereas states and interest groups can have some effect at the postlease stage through permitting processes, there is no existing means for states and interest groups to affect the auction itself.[18]

Postleasing: Revenue Sharing

Annual federal OCS income from 1979 to 1985 exceeded $5 billion; over $80 billion in OCS receipts were received since 1954. A revenue source of this size is important to federal budget planning, and, of course, is of interest to the states who view the distinction between state and federal lands as arbitrary.

The issue of state revenue sharing results in part from the fact that oil and gas deposits do not correspond to the arbitrary geographic line dividing state and federal lands. Because oil and gas migrate through rock, it is possible for a well on one side of an arbitrary line to drain an oil or gas pool that lies partially in land of another jurisdiction.[19] Several coastal states began lobbying and litigation to recover revenues tied to production and to onshore effects of leasing. The result was a compromise by the secretary of the Interior in 1978 that specified a "fair and equitable sharing"[20] of revenues between tracts in a zone, called the 8(g) zone, between three and six miles offshore. This ambiguous language quickly led to litigation by Texas, Louisiana, California, and Alaska; decisions at the district court level quickly extended the concern for revenue sharing beyond the issue of drainage.[21]

In addition to drainage, the court found that the sequence of leasing on adjoining state and federal offshore lands, which usually involve a state leasing first, enhanced bonus payments on the federal lands. Bonus payments occur because exploration on the earlier leases can generate information that reduces costs and focuses later exploration.

As the litigation proceeded through the courts, Congress stepped in to legislate what was meant by fair and equitable. In 1986 Congress passed amendments to the Outer Continental Shelf Lands Act, which gave the states involved 27 percent of all bonuses, royalties, and rents from tracts in the 8(g) zone.[22] This resulted in an immediate lump sum payment to the affected states of $1.5 billion from funds that had been held in escrow since 1978. Of the $1.5 billion, $640 million went to Louisiana, $420 million to Texas, and $340 million to California with smaller payments to four other coastal states. Smaller payments, between $20 and 60 million, will be paid from the account over the next 15 years, whereas payments from current production and newly leased areas will be made on a current basis.

Postleasing: Permitting and Commenting

States and interest groups have several opportunities to affect the course of exploration and development after a lease has been granted. These opportunities result from legislation such as the Outer Continental Shelf Lands Act, the Coastal Zone Management Act of 1972, the National Environmental Protection Act, and the special transboundary problems caused by oil and gas activities at sea. At the boundary, a constant flow of water and air causes transboundary environmental problems with the neighboring states containing the most sensitive environmental resources and the largest amount of development unrelated to activities on the OCS.

Though the OCS lands are an exclusive federal jurisdiction, activities such as fishing and oil and gas exploration require strong transportation and supply links with the neighboring states. Further, the revenue from fishing and tourism is subject to state taxation, whereas the revenue from oil and gas activity is taxed by the federal government and is exempt from most state taxation. States therefore have a monetary as well as a political interest in the promotion of particular activities.

Several transboundary problems have been resolved to date in different ways. In one issue relating to air quality, a court determined in 1979 that the Department of the Interior has primary jurisdiction over air quality rather than the Environmental Protection Agency (EPA).[23] However, neighboring states and local jurisdictions have direct control over the variety of environmental and construction permits associated with shore facilities. This permit control is a powerful bargaining chip for states since almost all offshore oil and gas activities are linked to shore facilities through pipelines that cross state waters and affect local jurisdictions through onshore processing and storage facilities and supply operations.

Neighboring states also have a direct say in the many permits required before production begins. As a result of the Coastal Zone Management Act of 1972 (CZMA), as amended in 1976, states developed individual coastal zone management plans that were to balance state and national interests. The Coastal Zone Management Act required that Minerals Management Service activities directly affecting the coastal zone should be done in a manner consistent with the state management plan, and, more specifically, that the postlease exploration, development, and production plans required by the Minerals Management Service be certified as consistent with that plan. Following litigation by California, this consistency provision was applied to the lease sales themselves, but in 1984 the Supreme Court ruled that lease sales as such were not subject to the consistency provision.[24] However, the explicit provisions in the Coastal Zone Management Act that applied to the postlease plans required by the Minerals Management Service do remain in effect and represent an important opportunity for state and interest groups to affect the details of oil and gas development. For instance, if a development and production plan is determined by the state to be inconsistent with its Coastal Zone Management Plan, the development and production plan must be revised and made consistent, or an appeal must be made to the secretary of Commerce to overturn the finding of inconsistency.

Table 2-4 represents the opportunities for states and interest groups to participate in the federal postleasing process. This does not include the permitting process required for onshore facilities. The permits and plans listed in the Table 2-4 highlight the formal opportunities for states and interest groups to comment and to directly affect offshore development.

While primary responsibility for the management of the OCS lies with the Minerals Management Service, many organizations provide comments and make decisions that affect the management of the OCS. Congressional revisions to OCSLA and the Coastal Zone Management Act have further involved interested parties in a politically oriented process.

COMBINING PURPOSE AND PRACTICE

The many institutions discussed in this chapter represent a diversity of purpose and power. As issues occur, alternative policies are analyzed by each group, positions are taken, and ultimately a policy is implemented.

Policies can be and are studied from a variety of viewpoints based on potential votes, on environmental philosophy, and on political philosophy, but such analyses are difficult to communicate. However, the policy analyses that dominate the formal arguments are the result of scientific and economic analysis. The next five chapters of this

Table 2-4
Participation in Postleasing Permits

Postlease Permits	Participation
Exploration Plan (oil spill contingency plan, environmental report).	State CZMA consistency, federal agencies, public comment.
Permit to drill.	Prior state consistency review, EPA discharge permit.
Development and Production Plan (oil spill contingency plan, environmental report).	Coordinate and consult with affected states and local governments, Coastal Zone consistency review, public comment.

Source: Five-Year Leasing Program, Appendix Q (Postlease Process, Regulatory Program and Performance Record).

book present examples (many of them based on economics) of quantitative policy analysis that have been or could be used to inform controversial issues in OCS management. This does not deny that feelings and other informal means of analysis have a part to play in the reality of OCS management, but that feelings are incomplete information on which to base management decisions.

Policy analyses do not, however, avoid the problem of defining the purpose of management. The many competing goals discussed in preceding sections remain. However, use of a natural or social scientific framework to study an issue often invokes the goal or purpose of that framework. For instance, petroleum produced on Alaska's North Slope cannot be shipped to other countries due to federal law, and furthermore, must be transported to domestic ports in American tankers.[25] An analysis of this policy requires trade-offs between U.S. merchant marine employment, possible differences in the safety records of the U.S. merchant marine compared to foreign shippers, the possibility of lower domestic petroleum prices, the importance of different sources of supply (North Slope of Alaska versus Mexico versus Venezuela versus Saudi Arabia), our balance of payments with Japan, and whether this issue has anything to do with national security. It takes an unusually clear statement of goals to choose among these trade-offs. One method of evaluating trade-offs in a policy analysis is to use an economic analysis that aggregates money paid and received by many people. In addition, an economically based analysis often adopts the goal of economic efficiency.

It is interesting to note that economic efficiency is nowhere mentioned in statutory legislation as a goal for the management of the OCS. However, many analysts, including some in the Minerals Management Service, do interpret economic efficiency as the goal of management.[26] It is useful to establish what economic efficiency is and what it is not, and why it is often used as a framework for analyzing the trade-offs between competing goals.

The goal of economic efficiency is achieved when no policy remains that could increase the income of one person or group by more than it decreases the income of everyone else. Economists have developed an elaborate theoretical structure that leads to this concept, but several of the key assumptions are important to recognize.

The first assumption is valuing policy actions by the amount people are willing to pay for the policy. Willingness to pay is often, though not always, captured by the market prices and quantities involved in a policy. For instance, if developing an oil field is expected to lead to profits of $10 million in today's dollars but results in a decline in fishery profits of $5 million in today's dollars based on the market prices of petroleum, fish, and the costs of producing the various items, then a basic analysis is that economic efficiency would be achieved by undertaking the oil development.

This basic analysis gets increasingly complicated (but in a way that can be analyzed) when market prices do not reflect the willingness of people to pay for an activity. For instance, people may be willing to pay to preserve a view from shore that does not include a drilling platform, which increases the value of not developing, just as people may be willing to pay to be less dependent on foreign sources of petroleum, which may increase the value of development. An extended analysis of an economically efficient action includes these measures of willingness to pay that may not be included in market prices.

The second issue is that policy actions on the OCS affect peoples lives and livelihoods over a period of time. Economic efficiency for policies that have effects in different time periods is achieved when the actions are compared in present value dollars, that is, when dollars in the future are compared to dollars in the present by discounting (applying a smaller weight) to the future dollars. Whereas this practice is widely accepted in business and by economists, the exact weight to attach to future dollars is often a source of debate.[27]

The final major issue in the use of economic efficiency is the role of the existing income distribution. In the theory of willingness to pay, that willingness, unsurprisingly, depends on one's income (including income from property holdings and other sources of wealth). As a result, the market prices and other measures of willingness to pay that an analyst might observe in the world are dependent on the current distribution of income. If wealth and income were distributed in a different way, we might observe different measures of willingness to pay. Petroleum and fish might be valued differently and, therefore, the economically efficient policy might be different.[28]

Why is this goal of economic efficiency, which is nowhere stated in the OCS laws passed by Congress, often used by analysts and sometimes accepted by the courts? The answer is the difficulty of determining how to trade off among the effects of a policy action. Economic efficiency provides an explicit and detailed framework that compares the effects of a policy on fisheries, tourism, the environment, and the oil industry. Uncertainty can still exist about the accuracy of measurement of these effects, but the framework for analyzing trade-offs is relatively clear. However, it remains a challenging analytical effort to compare the effects of a policy in present value terms even without considering the uncertainty of measurement and the effect of a different income distribution.[29]

Economic efficiency has therefore crept into the goals of management through its potential to frame the trade-offs involved in a policy action. It is a useful tool when carefully applied. There remain, however, substantial difficulties in measuring uncertain effects that occur in the future, effects that are not captured in market prices and in incorporating value judgments such as the fairness of the existing income distribution.

The difficulty associated with analyzing economic efficiency is one reason why economists and other analysts often advocate that more decisions be left to a market and fewer decisions left to governmental agencies. The concept made famous by Adam Smith, the invisible hand, is a statement that economic efficiency can be achieved by individual people who, although seeking their own self-interest in a market, inadvertently achieve the form of public interest defined as economic efficiency. The strongest advocates of leaving the market alone advocate privatization—the permanent distribution of public lands to private owners without concern for auctions, withdrawals, or the regulation of exploration and production. These advocates emphasize that the auction process is just a means for the federal government to redistribute money from oil companies to the public treasury. Advocates of privatization have a well-developed body of theory indicating that economic efficiency could indeed be achieved by privatization and that existing legislation might be sufficient to incorporate some factors originally ignored by a market, such as pollution.[30] These arguments, however, have served mainly to remind managers of the objectives they wish to achieve through government ownership and regulation in place of pursuing the goal of economic efficiency through a market process.

Part Two

Policy Analysis

Policy analysis, the topic of Chapters 3–6, requires the foundations laid in Part One, as well as a basis for analysis. The basis for good policy analysis is not the loudest voice, the saddest story or the glossiest presentation. Good policy analysis is based on explicit and testable theories, verifiable data, and the exchange of information. Bad policy analysis depends on implicit theories, assumed data, and closed door decisions.

These four chapters rely heavily, though not exclusively, on the framework of economics. In lesser degrees of detail, the chapters in Part Two also incorporate analytical frameworks drawn from geology, biology, engineering, chemistry, and physical oceanography. In particular, Chapter 3 addresses issues in resource estimation, whereas Chapter 4 (authored by Thomas Grigalunas and James Opaluch) investigates issues related to the environment with particular attention to oil spills. Chapter 5 presents analyses relating to income received by the federal government— the issue of receiving fair market value; Chapter 6 investigates issues at a more aggregate level related to the pace of leasing and to the accounting for income received from Outer Continental Shelf leasing.

Feelings, theory, and numerical analysis are three different ways of analyzing whether a policy creates movement toward achieving a congressional or organizational goal or economic efficiency. Statements of feelings abound in open hearings, written comments, and polls. These statements are important signals to politicians and policymakers. However, it is difficult to distinguish strength of feeling about an issue, and then to determine how feelings translate into actions such as voting, or purchasing oil, housing, or recreation.

Policy analysts trained as economists are often skeptical about these statements of feelings. If talk is cheap, analysts often look to determine if people are willing to let their pocketbooks speak and so rely on prices that people actually pay. It also happens that when people make statements of their feelings, there is an unstated theory and conclusion behind those feelings such as whether a market leads to overly rapid exploitation of natural resources or whether environmental costs are excessively large. Scientists frequently struggle to make these theories explicit and to investigate the validity of the conclusions.

An intermediate step between statements of feelings and numerical analysis of the effect of a policy is to make predictions based on a theory. Economics is one body of theory that leads to qualitative predictions such as whether the government will receive more money for a lease if more companies submit bids, or whether oil companies must expect to earn more from exploration when it is very uncertain that the area contains oil or gas. Economists, however, are trained to analyze movement toward economic efficiency. Sometimes a part of the efficiency analysis can be used to analyze movement toward other goals such as increasing exploration from lowering minimum bids or decreasing exploration from additional environmental constraints. Other disciplines, such as oceanography, geology, or biology also provide a framework that leads to predictions about the quantity of resources in an area, the effect of drilling, the effect of oil spills on biological communities, or other topics.

Different theories can, of course, lead to different conclusions about the effect of a policy. If the theories have been stated carefully, it is sometimes possible to find a logical flaw. More frequently, however, some experience with a policy will provide numerical data to determine the effect of a policy and the truthfulness of a theory. Numerical analysis is not a panacea, however. The data may not be accurate. In the planning stage, data are often informed guesses, and later they may reflect only gradually increasing knowledge. Different ways of analyzing the data might lead to different results. Finally, too little attention may be paid to clearly describing how the data measure movement toward achieving a goal.

Recognizing the limitations of quantitative policy analysis, the four chapters in Part Two focus on numerical studies of several important policy issues such as resource estimation and exploration, the effect of oil spills, the pace of leasing, and the effect of policies on the receipt of fair market value. Each chapter also contains additional descriptive information and the ways in which analytical information has been used in policy debates.

Chapter 3

Resources and Exploration

Resource estimation is a fundamental building block for policy analyses of the Outer Continental Shelf. Estimates of resources over large areas are used in macro policy debates about the importance of OCS energy sources, in preparing the five-year plan of auctions, and in guiding negotiations about deferrals. At the micro level, resource estimates are fundamental in determining the adequacy of a bid from the auction and, after production has begun, in determining the actual royalties that will be received by the government.

The first section of this chapter surveys the methods currently used to estimate resources on the OCS and the policy situations in which they are used. An example of a theoretical model of exploration by a firm is presented in the second section of this chapter; the final section presents a policy analysis of the role of information from drilling in frontier areas.

RESOURCE ESTIMATION

Several resource estimates for oil and gas resources on the OCS are presented in Table 1-1 (Chapter 1). Over time, and as planning progresses, the resource predictions become more location specific and more accurate. In part because Congress specified an interest in knowing the extent of OCS resources and because of the organizational history of the Minerals Management Service as part of the Geological Survey, the determination of future resource availability is a major commitment of personnel.[1]

A basic but important distinction is that oil and gas deposits become resources for human use only when price and cost conditions allow the profitable extraction of the oil and gas. Common terminology is that resources become reserves when the economic and technical conditions could allow extraction, and some discovery of the resource has taken place. Although further breakdowns into categories of reserves and resources are discussed in many sources,[2] much attention in policy debates is focused on the resources remaining to be discovered. These estimates of yet to be discovered resources combine three of the types of uncertainty discussed in Chapter 1: geological, technological, and economic uncertainty.

The usefulness of resource estimation in major policy debates is itself a topic of debate.[3] The degree of uncertainty, sometimes measured by a range of estimates, and the variety of ways of predicting resources lead to different estimates by different analysts. Whatever numbers emerge, however, interest groups interpret them in a way that is favorable to their cause. This is illustrated in Figure 3-1.[4]

Although Figure 3-1 is not meant to illustrate good policy analysis, it does illustrate the fact that more accurate resource data does not necessarily lead to a common policy conclusion. Resource data, or the analyses based on such data, requires interpretation in light of goals such as those discussed in the preceding chapter. The following pages provide a survey of the current methods used by the Minerals Management Service to estimate undiscovered resources and then a discussion of their application to policy.

Resource estimates for undiscovered resources in the Minerals Management Service are based on a simulation program called PRESTO (Probabilistic Resource Estimates, Offshore). It uses information provided by geologists in the Minerals Management Service on the distribution (the mean and a specified range) of the extent of a deposit, the thickness, the oil recovery factor, the probability of the deposit existing, and similar technical data.[5] The program then determines (from the distributions provided) a sample mean and range of the resources to be found provided that the resource exceeds a minimum size that allows for profitable production.

The current simulation method of determining the undiscovered resources differs from a consensus method of expert assessment used in 1981.[6] In fact, not only the method but also some important characteristics of the studies are different. For example, prices were different, state lands were excluded from studies in 1985 and 1989, several major discoveries in the years between the studies moved resources from the undiscovered to the discovered category, and disappointing exploration results in Alaska and the North Atlantic altered the probabilities of discovery in those regions. Nonetheless, public concern about "disappearing" offshore resources occurred when the resource estimates were compared. The studies indicated an apparent decline of 55 percent in undiscovered oil resources and a 44 percent decline in undiscovered gas resources between 1981 and 1985. An additional decline of 27 percent for oil and 18 percent for gas occurred between 1985 and 1989. These results are observable with greater detail in Table 3-1.[7]

The three sets of estimates from three reports provide a clear example of numerically different analyses of the expected availability of resources that have had an uncertain

Participants	When prices are high, participants favor	When prices are low, participants favor
Preservationists	Low estimates "High prices encourage overproduction."	Low estimates "Low prices encourage overconsumption."
Consumerists	Low estimates "There is just not much left to be found, no matter how high prices go."	High estimates "There is no need to raise prices."
Industrialists	High estimates "Significant new supplies can be found if prices are high."	Low estimates "Higher prices are needed to bring on more supplies."

Figure 3-1. Reserves and resources: What the estimates "say" to participants. (*Source:* A. Wildavsky and E. Tanenbaum, *The Politics of Mistrust: Estimating American Oil and Gas Resources,* copyright 1981 by Sage Publications. Reprinted by permission of Sage Publications, Inc.)

Table 3-1
Comparison of Recent OCS Resource Estimates

	Oil (billion barrels)			Gas (trillion cf)		
Area	1981	1985	1989	1981	1985	1989
Alaska	12.2	3.3	.9	64.6	13.9	0
Atlantic	5.4	.7	.3	23.7	12.3	4.5
Gulf of Mexico	6.2	6.0	5.6	68.2	59.6	64.3
Pacific	3.2	2.2	2.1	6.2	4.7	5.2
Total	27.0	12.2	8.9	162.7	90.5	74.0

Source: L. Cooke, "Estimates of Undiscovered, Economically Recoverable Oil and Gas Resources for the Outer Continental Shelf as of July 1984;" U.S.G.S. and M.M.S., "Estimates of Undiscovered Conventional Oil and Gas Resources in the United States," 1989.

effect on policy. When economic efficiency is the goal, information of the total resource to be extracted just signals entrepreneurs in other parts of the economy to prepare for the transition when prices rise sufficiently high that substitution to other sources occurs. However, when congressional or interest group goals are the objective, the variety of positions summarized in Figure 3-1 can occur. Finally, the large degree of uncertainty inherent in the methods may indicate that the several estimates are really, in a statistical sense, close to one another.

The degree of uncertainty of resource estimation for particular prospects has been studied in some detail. The accuracy of prospect specific resource estimates was studied in several articles in 1979 and 1980.[8] Uman, James, and Tomlinson[9] compared prelease, tract specific resource predictions prepared by the Geological Survey, which was then administering the program, with postdiscovery estimates of reserves. (The resource predictions were also based on the assumed existence of the resource.) The study concluded that the predicted values were between one-tenth and ten times the amount actually thought to exist after the tract began production. Davis and Harbaugh[10] criticized some of the methods and assumptions of Uman and his coauthors and implied that the degree of uncertainty may be even greater.

Clearly, advancing policy analysis to the point of numerical analysis of resource size for a specific tract did not do away with uncertainty or with controversy. Nor would it be unusual to find different analysts developing predictions that are different by a factor of two, the difference between the 1981 and the 1985 estimates of OCS resources. One expects the properties of the aggregate estimate of resources to be such that the mean is the sum of the means of the individual prospects, whereas the variance of the aggregate will exceed the sum of the variances of the individual prospects as positive correlation among prospects is typically assumed.[11]

Validation studies, however, often omit an important discussion. For instance, Uman and colleagues state, "The purpose of our report was to indicate the limitations of current procedures so that policy analysts might determine if this level of performance is good enough for the uses intended."[12] Unfortunately this determination is seldom made. In addition, issues that are particularly important in implementing policy analysis but are often overlooked in the rush to completion are whether a bias exists (whether resources are consistently over or underpredicted), and whether any use of the numerical analysis is sensitive to the size of the error (if it is not, one wonders why the numerical analysis is done).

One example of this omission is the comparison of high bids from lease auctions and the estimate of value computed by the Minerals Management Service. The range of error is so wide for the mean estimate of value (consider that price, cost, and exploration uncertainties are added to the uncertain size of the resource), that this information is ignored when the Minerals Management Service compares its point estimates of value with the high bid. Implicitly, a decision criteria that ignores the variation in possible outcomes accepts the assumption that the government and the firms are risk neutral, that is, they behave as if they are maximizing expected value.[13] The Minerals Management Service has modified its bid adequacy rules to include outside information into its rejection rules (see Chapter 6). However, room for additional research on the implications of this implicit decision criteria exists.

Ignoring the variation in estimates of value is an omission similar to using the mean as a measure of central tendency for the distribution of resources. As has been recently pointed out,[14] the use of the mean assumes that there is a quadratic loss function associated with over or under estimation of resources. Whereas this is a common assumption, it is also possible that errors in resource estimation do not have a symmetric, quadratic influence. For instance, Solow and Broadus[15] show that if a linear loss function for estimation error is used, then 11 out of 19 OCS planning areas currently estimated to have positive resources using the mean as a single measure are, in fact, best estimated as having zero resources.

The work by Solow and Broadus[16] reveals how implicit assumptions and standard practice can be important in policy analysis. Although the Minerals Management Service has been careful to report information about the range of possible resource estimates, the focus of the policy debate tends to be on "the" number most readily understood. In the case of resources, that number has been the mean of the estimated distribution. A better understanding of the policy weights given to over and under estimation of resources could have large policy implications. Best-guess estimates of zero resources, based on an alternative loss function, provide little evidence for auctions in those planning areas.

Finally, the point estimates of resources are heavily dependent on the prevailing economic conditions—such as oil and gas prices, the cost of materials, and labor and technology. These economic and technological factors affect resource estimates through interaction with the minimum economic field size and the minimum basin and area resource. The analysis of undiscovered resources requires that the resources be profitable to extract if found. That means that each field must be of sufficient size to justify development. In frontier areas, it means that the total resources must be sufficient to justify the development of a regional transportation system, typically a system of pipelines. These are nontrivial costs in areas such as Alaska or the North Atlantic. In practice, these minimum sizes are determined through auxiliary studies and programs and then used in PRESTO as a cutoff value.[17]

One way of addressing the role of economics in the assessment of undiscovered resources is to estimate the changing resource base as prices change. This produces a resource availability curve whose primary distinction from that of an economic supply curve is that no explicit statement is made that the quantity of resources will be discovered and produced.

A crude example of a resource availability curve for the Outer Continental Shelf can be inferred from estimates by the Minerals Management Service in the 1987–1992 five-year plan.[18] These estimates were based on resource predictions using various assumptions about the starting level of prices, the rate of growth of prices, the discount rate, and so on. Figure 3-2 plots the results when only the starting price for the analysis is changed. The lower portion of each bar indicates the estimated resource availability of both oil and gas in the Gulf of Mexico as the starting price increases from $14 to $29. The next segment adds the estimated resources from the Pacific region to those of the Gulf at each price. The resource availability curve moves further up as more regions are included in the estimates. The top line is the estimated total resource availability on the Outer Continental Shelf as a function of the starting price.

Clearly, substantial regional differences in the expected quantity of leasable resources exist depending on the starting price. The relatively mature Gulf of Mexico area has the largest estimate of available resources (in part due to the higher probability of finding a deposit in this area). If estimated prices more than double, from $14 to $29, then the available resources in the Gulf of Mexico would increase by only 17 percent. This contrasts most strongly with Alaska where, at $14 per barrel of oil, no available resources to be leased exist. However, if the starting price increased from $14 to $29, expected resources of over a billion barrels exist, 13 percent of the resources available at that price. The actual resource availability may, of course, differ substantially from the expected values as exploration results in more or fewer resources being discovered than expected at the time the estimates were prepared.

Several techniques for the quantitative analysis of resources have been illustrated. These techniques include the simulation methods of PRESTO, expert elicitation meth-

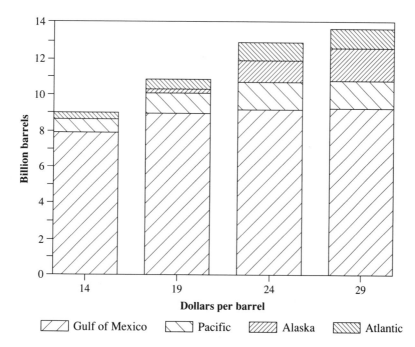

Figure 3-2. Resource availability curve: unleased, leasable resources. (*Source:* Minerals Management Service, *Five-Year Leasing Program,* 1987.)

ods used in earlier resource studies, and the regression and graphing techniques used to assess validation and resource availability. Whereas resource estimates are frequently discussed during policy debates, resource estimates have not been presented as critically affecting the debate. Yet, there are examples of policy actions based directly on resource estimates.

In the fall of 1985, Secretary Hodel was negotiating with the California congressional delegation about the number and location of tracts to be offered in an auction off the coast of northern California. The California delegation had been leaders in legislating the moratoria on leasing off parts of both the East and West coasts, and no auction had taken place off northern California since 1963. It appeared that Hodel and the delegation had reached agreement on approximately 150 tracts that could be offered for lease. However, Hodel withdrew from the preliminary agreement on the basis that further analysis indicated the poor resource potential of the 150 tracts.

Resource estimates are also important at the auction stage. Once auctions are held, it is not uncommon to find that bids have been received for which the Minerals Management Service finds that there is no resource potential according to the data available. On that basis, the bid of the company will be accepted as long as it exceeds the minimum bid (see Chapter 5 on fair market value).

Finally, the pace of leasing (see Chapter 6) is said to be strongly influenced by estimated value. These value estimates are approximately linear in the estimated resources. As a result, resource estimates greatly affect the planning for auctions by the Minerals Management Service. These examples of the influence of quantitative resource estimation on policy analysis should not take away from the fact that in many other cases, resource estimation does not have a clear-cut effect on a decision.

EXPLORATION: THE ROLE OF COST AND UNCERTAINTY

Discovery, reserve definition, and production begins with exploration. Exploration takes place in two major phases, each of which can be influenced by government policy. Prior to leasing, many companies either singly or together—without drilling—obtain scientific information on the resource potential of an area. Only after obtaining a lease and the appropriate permits, however, can a company drill to obtain information.[19] Furthermore, the typical diligence requirements of a lease require that drilling begin within five years of the lease date in shallow water, less than 400 meters deep, within eight years in waters between 400 and 900 meters deep, and within 10 years for water depths exceeding 900 meters.

Exploration is the way in which geologic and economic uncertainty is partially resolved. However, the oil industry is replete with examples of the difficulty of finding even major new areas of hydrocarbons. Paul Kobrin recently surveyed the exploration effort required to find three modern examples of new hydrocarbon provinces.[20] His examples are the North Sea, the North Slope of Alaska, and the Overthrust Belt in the western United States. Kobrin presents descriptions and quantitative evidence of the extensive exploration necessary to find the first discovery in these areas; see Table 3-2.

The rate of success is, of course, higher in existing production areas. Exploration is

Table 3-2
Exploraton in Three Provinces

Province	Dry Holes before Discovery	Years
North Sea (Oil-prone portion)	200+	~10
North Slope	~50	24
Overthrust Belt (Montana, Idaho, Wyoming, NE Utah)	~500	~80

Source: Paul Kobrin, "Finding Oil in the Arctic: Three Case Studies of Successful Exploration." American Petroleum Institute, Critique #021, April 1988.

still necessary, however, to actually determine whether oil and gas exist. This section presents data on the current rate of success and the timing of drilling in the most heavily developed offshore area, the Gulf of Mexico, and then presents a theory of exploration that highlights the role of cost and uncertainty in predicting the number of wells to be drilled. The predicted number of wells to be drilled has been viewed as an important policy variable since the number can change in a short period of time. In addition, the drilling of wells is an important step between issuing a lease and eventual production or abandonment of a tract.

In the Gulf of Mexico, where considerable development has occurred, less developed areas remain because tracts are never drilled but instead are returned to the government at the end of the primary term. Other tracts are drilled, often with multiple wells, but no discovery is made. Data on these issues are presented in Table 3-3 where the tracts leased, tracts drilled, and productive tracts are arranged by year. The most recent years indicate little drilling activity due to the time lags to plan drilling, obtain permits, and the depressed state of the oil market at the time the data were collected (as of March 1987).

Several issues are highlighted by data in the table. The first issue is the large increase in leasing activity starting in 1983 when a program called areawide leasing was instituted. (For the effect of this program on fair market value, see Chapter 5.) The second

Table 3-3
Exploration Drilling—Gulf of Mexico

Year	Tracts Leased	Tracts Drilled	Percent Drilled	Tracts Productive	Percent Productive
1979	171	151	88	91	53
1980	183	149	81	85	46
1981	258	155	60	84	33
1982	182	105	58	55	30
1983	1185	434	37	180	15
1984	814	154	19	51	6
1985	642	35	6	12	2
1986	142	1	1	0	0

Source: Department of the Interior, *Five–Year Leasing Program,* Appendix P, pp. P-16, P-17.

issue is that not all leased tracts are drilled, and that many tracts—often less than half in this thoroughly studied and well-developed area—are not productive. The decline over time in tracts drilled and tracts that are productive results in part because the later leases are still active (some being leased shortly before the data were collected). In contrast, the primary term for most of the earlier leases has expired, and the exploration record is complete. Whether the later leases will achieve the higher percentage of tracts drilled and found productive in earlier years remains to be seen.

A theoretical framework of searching for oil and gas illustrates the importance of cost and uncertainty while highlighting the effects of a company owning more leases over time and facing a cutback in the exploration budget. Even in the Gulf of Mexico, costs rise rapidly as the water depth increases, and the existence of oil and gas remains uncertain.

Theories abound about how companies choose their exploration plans—which tracts to explore, in what sequence, and when to stop. All of the theories require some assumption about the behavior of the company—the goals the company wants to achieve. The issues of interest can be highlighted by assuming that when companies plan their exploration, they are seeking the largest expected addition to their reserves for a given exploration budget.[21]

In order to determine what tracts owned by a company should be explored according to this objective, the objective must be rewritten with terms that include the cost of exploration and the probability of finding oil or gas on the site. Defining what is meant by expected reserve additions begins by defining the probability of discovering a deposit that in fact is on the site. The probability of discovering an existing deposit on a particular tract, call it tract j, increases as the cost spent on exploration increases, c_j, and as a cost adjustment factor, b_j, increases. The cost adjustment factor adjusts for the fact that spending a dollar in shallow water or a shallower drilling depth is more likely to lead to a discovery than a dollar spent in deeper water or drilling to a greater depth. In effect, more exploration occurs per dollar in some areas than in others, and this is reflected in different values of b_j. If the probability of not finding the deposit declines exponentially with increased expenditures, then it can be shown that the probability of discovery, given that the resource is on the site, is:

$$1 - \exp(-b_j c_j).$$

An additional probability, p_j, the probability that there is in fact oil and gas on the tract, must be included to determine the overall probability. Finally, the expected reserve addition is the overall probability times the size of the discovery, S_j. Thus the theory determines how a company will act if it attempts to maximize the expected amount of oil and gas discovered over all of the tracts it controls, subject to a limitation that it cannot spend more to explore than has been allocated in the corporate budget, C. Equation 1 is the formal statement of the problem:

$$\text{Max} \quad \Sigma\, S_j p_j (1 - \exp(-b_j c_j)) \tag{1}$$
$$\text{subject to: } \Sigma\, c_j = C$$
$$c_j > 0$$

The mathematical solution to this problem is an applicaton of nonlinear optimization.[22] The mathematical solution to equation 1 contains a condition that must be met if a solution exists. That condition (the first order condition) is:

$$b_j S_j p_j \exp(-b_j c_j) - \lambda \le 0 \quad j = 1 \ldots M \tag{2}$$

Equation 2 has a ready economic explanation that incorporates both cost and uncertainty. It requires that where exploration expenditures are positive (Equation 2 equals lambda (λ)) then lambda is equal to the marginal expected reserve additions per dollar spent on exploration, which is equal across all sites being explored, since lambda is a constant. For sites where expenditures are zero, i.e., $c_j = 0$, no exploration occurs, $b_j S_j p_j <$ lambda. This latter equation suggests that the firm explore tracts in a sequence from the highest to the lowest value in sequence of weighted, expected resource content, $b_j S_j p_j$, with the weight equal to the cost parameter, b_j. Tracts are more likely to be explored earlier the larger the exploration value per dollar, the larger the possible discovery, and the larger the probability of finding something. Alternatively, any one of the above variables or a combination of low exploration value per dollar, small discovery, or low probability can lead to not exploring a tract. All the factors must be taken into account.

If less money is available for exploration, then fewer tracts exceed the threshold of acceptability for drilling. In that sense, firms are responding to their own logic and that of the market when they decrease exploration activity in times of declining prices. In fact, changes in the oil and gas markets quickly become translated into changes in the budgets for exploration and development. Table 3-4 shows the changes in exploration budgets, falling almost 50 percent as a result of price declines of 1986.

Also, companies do not exhaustively explore a single tract before moving onto another location. Some tracts are leased and explored several times before a producible deposit is found or before all interest is lost. The way tracts move from the top, most likely to be explored, to the bottom of the priority list is that the probability of finding a deposit changes as information is obtained from drilling. This type of updating of information is addressed in more detail in the following section.

INFORMATION FROM DRILLING IN FRONTIER AREAS

This section presents an extended example of a quantitative policy analysis. The analysis determines whether the goal of expeditious exploration is hindered by the way

Table 3-4
Changes in Exploration Budgets (Billions of Dollars: 1985–1988)

	1985	1986	% change	1987	% change	1988	% change
Drilling-exploration (all U.S. projects)	26.6	13.6	−48.8	12.4	−8.8	14.4	16.1

Source: Oil and Gas Journal, February 23, 1987, February 22, 1988.

leases are currently granted.[23] If it is found that current regulations such as lease size are a major hindrance to exploration, then regulations could be interfering with several goals, among them economic efficiency and expeditious exploration. This policy issue is not idle speculation. In 1986 the Department of the Interior released for comment "Analysis of policies to encourage leasing, exploration, and production on the Outer Continental Shelf,"[24] and President Reagan proposed to Congress in 1987 that the minimum bid be lowered to encourage exploration in frontier areas of the OCS, an action later implemented by the secretary of the Interior.

One justification for the proposed changes was that individual companies do not capture all of the benefits from conducting exploration. In particular, the Department of the Interior argued that a company that explores in a frontier area inadvertently provides external information benefits to other firms. External benefit in this context means that something valuable—information on the probability of finding a deposit—is provided without charge to another firm. If the information is very valuable and if the information is publicly available, then an inefficiently low level of exploration could result. This section presents one framework for quantifying whether the external benefit is large or small.

In practice, drilling for oil and gas is the only way to determine if a deposit actually exists. Even a dry well—one that does not yield oil or gas—can provide useful information for siting other wells.[25] The broad geological forces that are believed to cause the creation and trapping of oil and gas in pockets extend beyond the administrative boundaries of a single tract and may cover thousands of square miles. Each tract, however, has its own geologic idiosyncrasies that differentiate it somewhat from these broad forces. The information gained by drilling reveals something about both the broad forces and about the site-specific details. One of the stated arguments for revenue sharing (see Chapter 2) deals with this same issue—oil and gas discovered first on state lands can increase the value of similar areas on federal land, whereas dry wells decrease value on similar land.

Recent economic theory has determined that the economically efficient level of exploration involves three effects: the change in value on the prospect explored, a depletion effect due to removing prospects from the exploration inventory, and the effect resulting from the change in information on the prospects yet to be explored.[26] Two of these effects may be external, and as a result, they depend on the distribution of property rights (who owns the related areas).

The importance for other property owners of information obtained by drilling on a particular prospect can be addressed in two ways. The first way is the effect of new information on the predicted level of exploration in the entire area when different patterns of private ownership are assumed. Patterns of private ownership are assumed to be either a separate owner for each prospect in the area or else a single owner for all of the prospects. The second way is the effect of a different measure of value that does not deduct taxes and royalty payments. This latter value is called the public value.[27]

If there is a single owner of the entire area, the stopping decision (which determines the level of exploration) incorporates the effect of information on all related drilling sites. The effect on related sites would be external effects in the case of many owners. This analysis of the stopping point based on new information emphasizes the range of exploration that can result depending on the probability of discovery; probability of discovery is again a key concept.

The Minerals Management Service is particularly concerned when there is no drilling interest in a frontier area, in other words when exploration "stops" prior to the first well being drilled. In a frontier area, one without existing production, the most valuable bit of information is whether commercial quantities of oil and gas exist in the area. The discovery of oil and gas is virtually impossible to keep secret; therefore, the problem is one of a public information externality. The company doing the exploring is effectively forced to reveal the information whether they wish to or not.

Whereas theoretically there are varying degrees of information such as the size of deposits, the frequency of discoveries, and geologic data that affect costs, these issues cannot be addressed with the data available. Furthermore, some of these factors may not be public knowledge and so would affect behavior in a different way. This is a frequent reality of policy analysis: the analysis provides new information, but the assumptions required for the analysis and the framing of the problem typically leave some issues debatable or unresolved. The creative part of policy analysis is the blending of issues, framework, and data to provide a convincing, if not iron-clad, argument that highlights the importance of specific issues.

The analysis will reveal that incorporating public information externalities has a small effect on the level of exploration in the area studied. Whereas the level of expected values are different when external effects are included for the prospect prior to the stopping point, the decision to drill is not significantly altered if one excludes all external effects. Based on government data, however, private decisions not to drill the earlier prospects may result in an expected loss as high as $729 million in the area studied—the shallow portion of the South Atlantic planning area.

The baseline for comparing the level of exploration with and without information is that in the area studied, six prospects out of 104 identified prospects have positive values. Table 3-5 presents the data used in the analysis for the shallow portion, less

Table 3-5
South Atlantic Shallow[a]

	MCRS MM-BBL	# of prospects[b]	Prob. of success[c]	Million 1987 $ Per Prospect		
				ATNPV	Royalty	Taxes
1	5	20	.013	−6.06	.09	−4.52
2	15	28	.014	−5.74	.28	−4.28
3	25	14	.014	−5.43	.47	−4.02
4	65	18	.016	−10.33	1.03	−7.51
5	150	6	.02	−7.19	2.97	−4.94
6	300	12	.028	−6.56	6.80	−3.42
7	550	4	.037	5.05	17.67	6.90
8	1250	2	.074	80.85	80.22	72.63

[a] Parameters: area risk (P): .75; discount rate: .08; price of oil: $19 per barrel in 1984 dollars; real change in the price of oil per year: 1 percent.

[b] The Department of the Interior placed each of the 104 identified unleased prospects into one of eight resource size categories, depending upon the mean conditional resource level, without changing the total expected resource amount in the area.

[c] The success probabilities were estimated by Interior for each resource size category by using regression techniques from the original prospect probabilities. This approach maintained the proprietary nature of the data without materially affecting the results.

Source: Minerals Management Service.

than 200 meters depth, of the South Atlantic planning area. In the absence of information, these six prospects are expected to be drilled. This sequence of exploration occurs regardless of the pattern of ownership since there are no gains or losses from new information.

The data indicate the existence of 104 prospects, though only those with a mean conditional resource size (MCRS) greater than or equal to 550 million barrels were determined to have a positive after-tax net present value (ATNPV). For prospects in the different size classes of this area, increasing probabilities of success on each prospect were found to be associated with increasing mean conditional (on hydrocarbon existence) resource size. However, the area in aggregate was thought to have a .75 probability of being devoid of hydrocarbons as measured by the area risk, P.

The effect of information to be analyzed depends on revising the probabilities of the area risk and its effect on prospect specific probabilities of success. Several quantitative tools are used in the analysis. The basic framework is that of Bayesian decision analysis. In that framework, new information updates probabilities based on Bayes formula, a well-known result in statistics. The revised probabilities are included in a decision tree where a company faces a sequence of decisions, whether or not to drill, as new information is provided.

The numerical analysis requires that the value of a prospect j be recomputed as a result of the information obtained when prospect i, a different prospect, is drilled. This recomputation also requires a quantitative model for estimating the value of a prospect when key geologic or economic factors change. The Minerals Management Service has developed such a model, which uses input data from geologists, engineers, and economists, and then uses a series of equations written in a spreadsheet format.[28]

In this analysis, the information that is obtained is whether oil and gas exist in a new area. As an example, consider drilling in new areas of the South Atlantic or in Alaska. The way that drilling on prospect i changes the value of prospect j is by changing the probability that oil and gas will be found on prospect j. Because of their broad geologic similarity, information from prospect i provides partial, but not complete, information about prospect j. If the value of prospect j is denoted by V_j, the probability of discovery on prospect j by G_j, the act of drilling on prospect i by D_i, the probability that the entire area is dry by P, and the change in each of these by delta, then:

$$\frac{\Delta V_j}{\Delta D_i} = \frac{\Delta V_j}{\Delta G_j} \frac{\Delta G_j}{\Delta P} \frac{\Delta P}{\Delta D_i}$$

This equation represents a chain effect. Drilling on prospect i, which is either successful or not, changes the probability of success on prospect j. This change in probability then changes the value for prospect j.[29]

It can be surprising how rapidly bad results, drilling and finding nothing, can in theory decrease the probability of finding oil and gas on a related prospect. An example from the data is presented in Figure 3-3, which shows that if there is a sequence of dry prospects, the probability of failure on prospect j approaches .95 after 8 dry prospects. Note that dry prospects in other areas never do drive the probability of success on prospect j to zero, and new interpretations of data often lead geologists to completely revise their probabilities in any event.

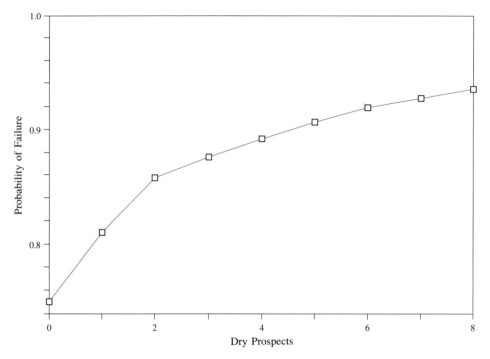

Figure 3-3. The probability of failure.

Exploration Effort: Private Value

The decision of whether or not to drill on a particular prospect depends on the outcome of prior drilling. For a privately owned company, a typical assumption is that it will choose not to explore if the net income after taxes that is expected falls below zero. This expected income for a company that owns only one prospect is shown in column 1 of Table 3-6. The data in that table show how sequential dry prospects cause a decline in value (a single discovery would lead to many more prospects being drilled which is

Table 3-6
Conditional Net Benefits Failure Path (million 1987 dollars)

Prospect i	Private Net Benefits			Social Net Benefits		
	Direct V_j (1)	External (2)	Total Value (3)	Direct V_j (4)	External (5)	Total Value (6)
1	80.85	129.71	210.56	233.70	495.56	729.26
2	58.71	24.60	83.31	172.57	162.35	334.92
3	−2.11	7.27	5.16	7.62	74.96	82.58
4	−3.33	4.37	1.04	3.83	44.32	48.15
5	−4.41	3.01	−1.40	.51	23.69	24.20
6	−5.36	1.87	−3.49	−2.42	10.89	8.47
7	−10.30	1.13	−9.17	−16.90	6.54	−10.36

not a problem of a slow rate of exploration). In column 1, the owner of the third prospect in this area would choose not to explore that prospect if there had been two dry prospects previously because the expected value at that point is negative. This would be the stopping point of exploration when each prospect is owned by a different company and no owner receives the value of information that they generate.

When there is a sole owner of the area, all of the expected values including the external effect must be incorporated and added to the direct expected values from column 1. Because the expected external effect can be shown to be greater than or equal to zero, a sole owner will drill at least as many wells as when each prospect is owned by a different owner, the case of separate ownership. The external effect is reported in column 2 of Table 3-6 and the total effect, the sum of columns 1 and 2, is shown in column 3. The level of exploration effort is increased to four prospects when all external effects are incorporated. Even a sole owner would not wish to drill more than four dry prospects in sequence as prospect five has a total expected value of -1.40 million.

The difference in exploration effort with many or only one private owner is thus only two prospects. Three key factors account for this minor divergence when the externality of information is included. One is that the expected external effect becomes progressively smaller as the probability of the area having oil or gas decreases. In other words, the information of a dry prospect has increased the chance that the area is dry. The second factor is that as the best prospects are removed from the inventory and drilled, there are fewer undrilled prospects associated with an external effect. This is sometimes called a depletion effect because there is a finite number of oil and gas prospects to be drilled. There are fewer prospects whose value is unknown due to a lack of exploration. Finally, because the prospects are drilled in the order of their expected value, at the stopping point there are no potentially positive valued wells to be incorporated into the expected value of the current well.

Exploration Effort: Public Value

A second comparison can be made in the levels of exploration effort along a sequence of dry wells when a different value, public value, is used for each prospect. Public value uses net income before government payments to define value analogously to the private value in the first three columns. Public value is presented in columns 4 through 6 in Table 3-6. If there were many owners evaluating the prospects using social values (corresponding to the case of no tax or royalty payments) then the level of exploration effort would be five prospects (determined by the prospect at which the value goes negative in column 4). Including all external effects would result in the sixth dry prospect becoming the maximum level of exploration effort in a sequence of failures. This is because the expected value goes from positive to negative between the sixth and the seventh wells when external effects are included as indicated in column 6.

In sum, the results indicate that distortions caused by government payments *and* the failure to include external effects cause a conditional loss of $82.58 million if only two prospects are explored instead of six prospects. This loss is conditional on the probabilistic outcomes actually evolving down this dry path of sequentially dry wells. The expected loss, which takes into account the probability of following the dry path, is $72.18 million.

These quantitative results are subject to different interpretations just as is the level of the resource presented in Figure 3-1. Organizations or individuals advocating greater exploration might say that three times as many prospects would be drilled with a sole owner who did not have to make payments to the government. A more valid approach is to ask what is the value of the different exploration sequences in the context of the cost of regulations and other distortions in the economy. If the number is accurate, a $72.18 million expected loss is, in fact, a small social loss in the context of a program generating billions of dollars per year in total value.

The Value of Prospects Drilled

The external effect does significantly increase the value of the prospects that are drilled. This can be noted by observing the relative size of column 2 compared to column 1 and column 5 compared to column 4 for prospects one and two. From a policy point of view, however, the higher expected values do *not* alter the decision to drill. Private industry is expected to have sufficient incentive to drill the earlier prospects in any case because their value exceeds zero.

However, the divergence in actuality between government measures of value and privately developed measures of value can significantly affect the interpretation of the first prospects. Suppose that industry is not interested in drilling in this frontier area. The values measured by industry must then either be less than zero or, due to risk aversion or some budget constraint, fail to pass some augmented threshold for drilling. In essence, industry may decide not to drill the first well. If the government estimate is correct, this decision results in a loss of potential net benefits of $729.26 million (the total value associated with prospect 1).

Differences of this magnitude, a $72-million loss or a $729-million loss or somewhere in between, leave a substantial degree of uncertainty about what policy would, in fact, create movement toward the goals of expeditious exploration and economic efficiency. This can be emphasized by considering several further interpretations of the results. First, an analysis using numbers generated by industry may be inappropriate due to different values for key variables in the analysis. Examples might be the discount rate, price, and resource distributions, with the result that expected social losses of not drilling the first well are, in fact, equal to the government estimate of the reported social value of that well. (Recall that we have not yet even considered other goals such as balancing environmental impacts with development.) Alternatively, if the government figures are inappropriate, the expected value of the first well could be revised downward. For example, the expected social value, including the externalities, falls to zero if a starting price of $13.50 per barrel in 1987 dollars is assumed.

The policy implications of this research are very different depending on the validity of the source of the numbers. If the government model is valid, there is a relatively large loss from not drilling the first few prospects, though the major source of the loss would not be the external effect. If no drilling is observed, policymakers may ask what mechanisms are available that result in achieving economic efficiency or other goals. This turns out to be a difficult issue in the choice of policy instruments and bureaucratic implementation. One possibility is to alter royalty payments or to grant a monetary prize for the first discovery. There are complications in these approaches, however, when many interest groups do not wish to see any further exploration, and the implicit or explicit payment would have to be quite large to encourage a firm to drill.

Alternatively, if the model is not valid, the loss from not drilling the first few prospects may be much smaller (though the analytical framework suggests that the distribution of property rights remains a relatively minor issue). Decision makers may wish to consider evidence of the validity of the value estimates as discussed earlier in this chapter.[30]

The management of the mineral resources of the OCS has been shown in this chapter to be intimately intertwined with estimates of the amount of resources that exist and with processes that lead to their discovery. Congress has spelled out its goals, but the Minerals Management Service must interpret them while relying on private companies to undertake the actual exploration, and, if successful, production. Geology, costs, prices, and the value of information all affect the outcome, and all these elements are subject to uncertainty.

Chapter 4

The Environment†

Environmental issues have been an important and contentious element in OCS oil and gas leasing policy since at least the time of the 1969 Santa Barbara production platform oil spill. Debate concerning the possible environmental effects of OCS oil development moved from being a primarily regional concern to a national issue in the early 1970s when the U.S. Department of the Interior began to consider actively leasing in frontier areas throughout the United States. Much has been learned about the effects of oil and oil development on the marine and coastal environments in the years since the Santa Barbara spill, and many important changes have occurred in industry practices and in the regulatory environment. Nonetheless, environmental risk and perceptions of risk continue to play a prominent role in OCS oil and gas leasing policy.

The importance and controversial nature of the environmental issues associated with the leasing of OCS lands are reflected in the many court suits that have accompanied proposed lease sales and other OCS activities and in the actions of congressional committees with respect to establishing moratoria and deferrals for leasing various OCS areas. Another indication of the significance and contentiousness attached to the environmental aspects of OCS leasing is provided by the many pieces of legislation that impose environmental restrictions on OCS oil and gas development. Major federal laws include the National Environmental Policy Act of 1969, the Marine Mammals and Protection Act of 1972, the Endangered Species Act, the Coastal Zone Management Act of 1972, and the Outer Continental Shelf Lands Act Amendments of 1978. These acts establish the policy and legal setting for environmental concerns within which the Minerals Management Service must operate.

The Outer Continental Shelf Lands Act (OCSLA) Amendments of 1978 were intended to expedite exploration and development of the OCS to achieve national economic objectives, while at the same time balancing the pursuit of these national objectives with protection of the human and coastal environments. The Act reflects recognition by the congressional and executive branches of government that the pursuit of the national benefits from OCS oil and gas production may impose environmental costs, and the potential trade-offs between national economic objectives and environmental objectives must be considered. However, neither the OCS Lands Act nor the

---

†This chapter is written by Thomas A. Grigalunas and James J. Opaluch.

policy statements by the Department of the Interior assign specific weights to individual environmental risks, and, indeed, it is extremely difficult to quantify the types of environmental risks involved.

In addition to national considerations, the OCSLA requires that regional equity must also be sought. Hence, trade-offs from leasing OCS oil and gas resources must be considered, not only for the nation as a whole but also for each region. Clearly, these legal requirements are a formidable burden on the Minerals Management Service with respect to the analyses that must be carried out to support an OCS oil and gas leasing program, the generation of data to support these analyses, and the integration of these analyses in a framework useful for the policymaking process.

Whereas many organizations, including states, industry, and environmental groups, undertake environmental studies of the OCS, the Minerals Management Service has a substantial organization devoted to environmental studies. An environmental assessment division exists at the headquarters level as well as at regional levels. In 1986 the Minerals Management Service spent approximately $25 million on its environmental studies program, and a total of $435 million has been spent since 1973. In addition, the Minerals Management Service has organized a group of outside experts into an OCS Scientific Committee, which advises the director of the Minerals Management Service on the environmental studies program.

This chapter explores two specific ways in which economics has been used to incorporate environmental issues into the 1987–1992 five-year plan of the Minerals Management Service and to determine the effect of environmental issues on the auction process. The first way quantifies environmental risks and costs, and the second way assesses the effect of liability for damage to the environment on the behavior of oil and gas firms.

OCS oil and gas development poses several types of potential environmental risks. These risks include those resulting from large and small oil spills as well as nonspill effects. The latter include potential air quality effects, loss of wetlands, and physical conflicts with commercial fishing. The potential environmental costs are referred to as "external costs" because they are costs imposed on others outside of the firm, unless compensated for by the responsible company.

If all of the external environmental costs of OCS development could be quantified with precision and if only economic efficiency mattered, a benefit-cost approach would provide relatively clear-cut economic guidance for developing a national lease schedule. Defined in this way, economic efficiency can be considered as a goal that can be used for policy analysis to rank OCS areas and to examine trade-offs that can arise in developing an OCS oil and gas leasing program.

However, in most cases it is extremely difficult to quantify environmental costs with any precision. These difficulties stem from the need to consider interdependencies among a range of factors when assessing possible future environmental costs. For example, to estimate the risk from spills, it is necessary to estimate the magnitude of resources to be produced (which depends upon the price of oil and other factors as discussed in Chapter 3) and on annual production, since the risk of spills depends upon the volume produced or transported. Further, the transport mode(s) must be identified, since tankers have different spill safety records than pipelines, and the size of spills differs between the two transport alternatives. The likely fate of spilled oil is important since spills that strike a beach, for example, will impose different costs than spills that go out to sea.

Hence the chance that a spill will come ashore also is important in assessing risks. Finally, to measure environmental effects in monetary terms given the fate and effects of the oil, appropriate valuation approaches must be used to establish costs. Some spill-related costs (clean-up costs, for example) are relatively straightforward to measure through the market; other costs (losses in beach use value or loss of the services provided by wetlands, for example) are no less real but are much more difficult to quantify since they are not generally bought and sold in markets.

Quantification of environmental risks and costs is particularly difficult when attempting to address the broad trade-offs facing decision makers at the planning area level—the level of analysis appropriate for developing a national leasing schedule. In some cases, potential societal costs—for example, effects of substantial OCS development on the lifestyle of native communities—cannot meaningfully be assessed using economic considerations.[1]

Despite the many difficulties involved with measuring environmental costs in monetary terms, efforts to improve analysis on this topic are important. Improved analysis is important because to the extent costs can be stated in dollars, the various risks from oil and gas development in an OCS planning area can be viewed in relative terms, and risks among areas can be put on some comparative basis. More fundamentally, stating risks in dollars allows for a broad comparison with the benefits of development, which are measured in dollars.

The first section of this chapter reviews two approaches for quantifying environmental costs from OCS development. These approaches represent an initial effort to measure environmental costs and thereby put such costs in some perspective. For example, the results can be used to gain an appreciation of the relative costs across regions and can be used to help rank areas by providing at least a first-order measure of potential net social value (development benefits minus environmental costs). The challenge is to develop a conceptual framework that can bring together important but disparate environmental factors that determine environmental costs in a way that can contribute to the policy process while at the same time recognizing the many uncertainties involved with any such analysis. Nonquantifiable effects and other social goals, as is always the case, become additional factors to be considered among the multiple goals of a leasing program. One approach uses a survey of case studies of oil spills and other external costs to estimate the unit costs of effects from OCS exploration and production. The second approach is a more integrated framework that directly links the fate of spilled oil with its effects. Whereas the newly developed fate and effects model is a more consistent approach, both models are discussed because of the importance of the unit cost model as background and its importance in the five-year plan.

The third section of this chapter deals with a second contribution economics can make to environmental analysis for OCS management. That contribution is in designing policies to assess firms for liability for damages and in evaluating the potential effectiveness of liability rules that apply to OCS oil and gas. For example, under the OCSLA and related federal legislation, polluters are strictly liable for compensation for damages from OCS-related oil spills. These liability rules were intended to achieve the equity goal of compensating those who experience losses from OCS activities. However, as an unintended side effect, the liability provisions of the federal laws covering the OCS also may serve as a "tax" on pollution and, thereby, provide economic incentives for firms to avoid pollution activities.

ASSESSING ENVIRONMENTAL RISKS AND COSTS

The external costs that can arise from OCS oil and gas development include those that result from these major sources: oil spills in the marine environment, and nonspill costs, which include physical conflicts among competing marine resource uses and other adverse coastal impacts.[2] Oil spill costs include, for example, possible losses to commercial fisheries and losses to beach users. Clean-up costs also are included as part of the cost of an oil spill. Examples of nonspill costs include losses to commercial fishers resulting from damages to fishing gear from contact with OCS oil and gas-related obstructions or debris and possible losses to commercial fishing, if OCS facilities physically prevent commercial fishing on sections of fishing grounds. Adverse coastal impacts result, for example, when onshore development leads to loss of coastal wetlands and the services they provide, such as habitat for harvested fish species.

To assist in the evaluation of the consequences of OCS development, two alternative measures of environmental costs relevant for policy analysis of OCS oil and gas are provided: social costs and regional costs. Social costs, generally speaking, include all spill and nonspill costs and are measured from the viewpoint of the nation as a whole. For example, the environmental costs that result when oil is produced in the Navarin Basin on the Alaskan OCS and spilled off the coast of Washington or Oregon en route to a refinery is attributed to the Navarin Basin as a social cost of OCS operations in that area. This reflects the fact that these environmental costs would not have occurred in the absence of Navarin Basin production.

In principle, the social cost for a given area is the aggregate amount that individuals throughout the country would be willing to pay to avoid the adverse effects of additional OCS development. The measurement of social costs is important because it is on the same accounting footing as the estimates of the benefits from OCS production. Hence, the measurement of social costs allows for an evaluation of net benefits (development benefits minus social costs) of OCS oil and gas resources.

The second measure of environmental costs used in this analysis is referred to as regional costs. Regional costs can be viewed as the total amount that residents of a given OCS planning area would be willing to pay to avoid the increase in OCS activity. These are the spill and nonspill costs measured from the perspective of the residents of a particular area as a result of production in *all* OCS areas.[3] For example, the regional costs from OCS development for residents of the Washington/Oregon area would include those costs caused by production off their shoreline. Regional costs for the area also would encompass costs resulting from oil spills from tankers carrying Alaskan OCS oil production through coastal waters. In contrast with social costs, the concept of regional costs specifically recognizes the geographic distribution of the costs of OCS operations.

Thus both of these measures of costs are critical for the decision-making process established by the OCSLA. Social costs are needed in order to include environmental costs in the evaluation of economic efficiency. Regional costs are needed in order to address distributional issues related to the "equitable sharing" goal of Sec. 18 (2) (B).

In summary, social costs (SC) and regional costs (RC) can be defined as:

SC = (Oil Spill Costs + Nonspill Costs)
   − (Reduced Oil Spill Costs from Lower Imports)

RC = (Oil Spill Costs Within the Region From Production in All Areas)
+ (Nonspill Costs Within the Region)
− (Reduced Oil Spill Costs within the Region from Lower Imports)
− (Compensation for Oil Spill and Fishermen Gear Losses)

The two most unusual factors in these definitions are the reduced oil spill costs from lower imports and the compensation for oil spill and fishermen gear losses. The first factor is necessary so that costs are incremental costs. In particular, the incremental social and regional costs must be measured net of the environmental costs of using imported oil. This is because additional oil and gas production from the OCS will replace imported oil resulting in lower risk of oil spills from foreign tankers. The second factor results from the possibility of compensation for some environmental costs. Under the OCSLA, a Fishermen's Contingency Fund was established to compensate commercial fishermen for losses of gear attributable to OCS facilities or debris (an item included in the financial analysis of Chapter 6). The OCSLA also established strict liability for removal costs and for a variety of damages from oil spills including losses to coastal businesses and damages to natural resources. Although compensation does not change the magnitude of social costs, it redistributes the loss by making oil companies liable to those in a region who suffer certain types of losses as a result of OCS operations. Hence, an effective compensation scheme reduces regional costs.

Measuring Social and Regional Costs

A wide range of economic, geologic, technologic, and environmental factors influence the potential magnitude and composition of the environmental costs of OCS oil and gas development. These include the estimated economically recoverable resources in each area, the time profile of production, transportation modes (pipelines and tankers) and routes to petroleum refineries, the chance of oil spills and their expected size, the expected fate of spills, and the marine and coastal resources at risk from spills. Nonoil spill costs, such as preemption of commercial fishing grounds, conflicts between OCS equipment and fishing gear, and disruption of wetlands, depend upon such factors as the aggregate scale of development in an area and the value of the resources affected.

The state of the art of environmental cost estimation does not allow for a precise estimate of potential environmental costs of OCS development. However, a substantial literature that quantifies many of the individual costs has begun to emerge, and recent interdisciplinary studies provide considerable insight into the range of some environmental costs of OCS oil spills.[4] Nonetheless, given the geographic scope of the problem, the aggregate planning-area level of the analysis, the many uncertainties associated with the possible location/timing of spills or other environmental effects within an area, and the difficulties inherent in quantifying environmental costs of pollution incidents, it is not possible to arrive at a precise measure of the true cost of OCS activity. However, it is possible to provide a useful insight into the magnitude of these costs. This is done by adopting a conservative stance whereby overstated estimates are used to represent the unit cost of OCS activities. In addition, sensitivity analyses are used to determine how the magnitude of total environmental costs for an area would vary in response to a change in one or more of the individual environmental costs considered.

The approach used to capture the interplay of the myriad of factors that determine the environmental costs of OCS development is illustrated in Figure 4-1 and outlined below. The point of departure for the analysis of social and regional costs is provided by the estimates of the expected, economically leasable oil and gas resources for each of the 22 OCS planning areas believed to contain resources for the range of oil prices considered. These resource estimates are a further refinement of the 1985 estimates presented in Table 3-1; a map of the planning areas is presented in Figure 6-3.

To reflect the uncertainty surrounding future oil prices, prices ranging from $14 to $29 per barrel were considered, though only the results from the low price scenario are

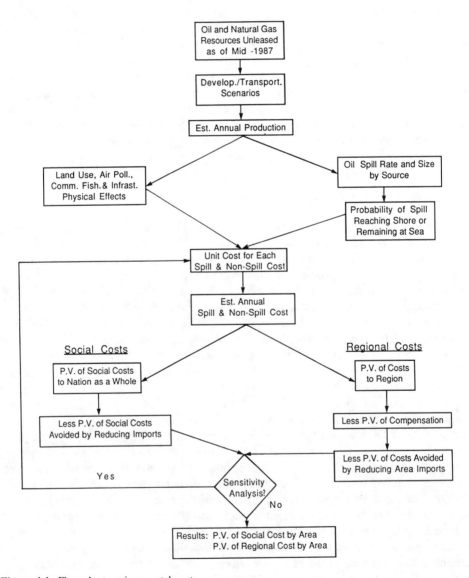

Figure 4-1. Flow chart-environmental costs.

reported here.[5] The DOI estimates of economically leasable resources in the $14 per barrel case are presented in Table 4-1. Resource estimates for each area are stated in billions of barrels of oil equivalent (BBOE). To construct this aggregate measure of resources, natural gas is converted to its oil equivalent (on a Btu basis) and then added to the estimated crude oil resources. Table 4-1 also indicates the likely transportation mode(s) for oil and the probability that oil that is spilled will strike land.

Given this range of resource estimates, a detailed assessment of the development strategy and the technology for developing and producing these resources, it is possible to obtain an estimate of annual production for each OCS area for each price-resource case. This effort uses available estimates for each OCS area of the representative time to production and to peak production, and estimates of peak production as a percent of total production.[6]

Table 4-1
Unleased Resources, Transportation Modes, and Probability of Spill Reaching Shore[a]

Summary of unleased resources as of mid-1987, transportation mode and probability of spills reaching land for each planning area for $14 oil starting price

| Planning area | Expected leasable resources | | | Transportation mode | Probability of spill reaching land |
	Oil (BBOE)	Gas (BBOE)	Total (BBOE)		
Central Gulf of Mexico	1.60	2.33	3.93	P/T	.772
Western Gulf of Mexico	1.08	2.71	3.79	P/T	.728
Southern California	0.28	0.10	0.38	P/T	.500
Central California	0.09	0.03	0.12	P/T	.444
Navarin Basin	0.00	0.00	0.00	P/T	.150
Beaufort Sea	0.00	0.00	0.00	P/T	.231
Chukchi Sea	0.00	0.00	0.00	P/T	.225
Northern California	0.10	0.08	0.18	P/T	.444
East Gulf of Mexico	0.09	0.09	0.18	P/T	.432
South Atlantic	0.05	0.20	0.25	P/T	.039
St. George Basin	0.00	0.00	0.00	P/T	.144
Mid-Atlantic	0.03	0.08	0.11	P	.106
North Atlantic	0.00	0.01	0.01	P/T	.100
North Aleutian	0.00	0.00	0.00	P/T	.342
Washington-Oregon	0.01	0.04	0.05	T	.650[a]
Gulf of Alaska	0.00	0.00	0.00	P/T	.851
Norton Basin	0.00	0.00	0.00	P/T	.250
Kodiak	0.00	0.00	0.00	P/T	.545
Straits of Florida	0.00	0.00	0.00	P	NA
Hope Basin	0.00	0.00	0.00	P	NA
Shumagin	0.00	0.00	0.00	P	NA
Cook Inlet	0.00	0.00	0.00	P/T	
Rest of Alaska	----	neg.	----	NA	NA

[a] Because no estimate has yet been made of the chance that an oil spill in the Washington-Oregon area would strike land, we have used the average of the indicated probabilities for the nearest two OCS areas, No. California and the Gulf of Alaska, as the estimate for the Washington-Oregon OCS planning area.

Risk of Oil Spills

Once annual production has been estimated for each area and resource case, the expected amount of crude oil spillage from large (at least 1,000 barrels or 42,000 gallons) and small spills (less than 1,000 barrels) can be estimated. Large spills from OCS production are rare. Only one OCS related spill greater than 1,000 barrels has occurred since 1981; and only four spills (totalling 27,000 barrels) were reported over the recent period 1975 through 1988.[7] Nonetheless, large spills can be expected to occur periodically and can cause large damages, particularly if they strike sensitive marine resources and reach the shoreline.[8]

Spill rates are typically expressed as the estimated number of spills per billion barrels of oil produced or handled. The spill rates used adopt the results of Lanfear and Amstutz.[9] The Lanfear and Amstutz statistical analysis of spill rates indicates that one platform and 1.6 pipeline spills greater than 1,000 barrels are expected to occur per billion barrels of oil produced or handled. Note that spill rates have declined over time as indicated in Figure 4.2. In fact, recent estimates are lower than the results used in this chapter to estimate oil spill costs.

To estimate tanker spill rates, Lanfear and Amstutz used worldwide vessel data for the period 1974–1980. They found that one would expect 1.3 large oil spills from tankers per billion barrels of oil transported, with 0.9 occurring at sea and 0.4 in port. A comparable vessel spill rate has not been estimated for U.S. waters; hence this worldwide rate for vessels is used for domestic vessels carrying OCS oil and for foreign tankers that would deliver imported oil in the absence of OCS production.

In addition to large spills, 395 small spills are expected to occur per billion barrels of oil produced on the OCS, the vast majority of which will be less than 50 barrels. This information is based on reported experience from 1964 to 1983 in the Gulf of Mexico, where over 90 percent of domestic OCS oil production has occurred.[10]

Regarding the estimated size of large spills, a central issue is how to represent the size of the "typical" large spill. Spill statistics are dominated by a few large spills, as is made clear by the following:

- The 77,000 barrels reported by Lanfear and Amstutz as spilled in the 1969 Santa Barbara incident accounts for 35 percent of all oil spilled in the 12 large platform spills since 1964.
- The 1967 West Delta pipeline spill of 160,638 barrels represents 77 percent of all oil spilled in the eight large pipeline spills since 1967.
- The five largest tanker spills comprise almost one-fourth (24 percent) of all oil lost in large spills in the period 1957–85; one accident alone, the *Amoco Cadiz* supertanker incident, accounts for about 6 percent of all tanker oil spillage throughout the world during the 18-year period.

The arithmetic mean of past spills in the United States is used to measure the size of the average large spill for platforms and pipelines.[11] The mean large oil spill size is 18,378 barrels for platform spills and 25,937 barrels for pipeline spills. For tanker incidents, the mean large spill size is 14,707 barrels for domestic vessels that would be used to move OCS oil and 20,769 barrels for foreign tankers used to carry imported oil. Thus, the average large spill from foreign tankers is 40 percent greater than large

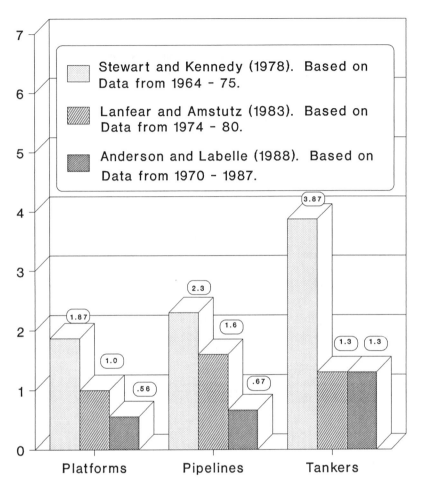

Figure 4.2. Trends in large spill ($\geq$ 1,000 BBL) rates from tankers and OCS platforms and pipelines (Spills per billion BBL oil handled). (*Sources:* K. J. Lanfear and D. E. Amstutz, "A Reexamination of Occurrence Rates for Accidental Oil Spills on the U.S. OCS," 1983 Oil Spill Conference; C. M. Anderson and R. P. Labelle, "Update of Occurrence Rates for Accidental Oil Spills on the U.S. OCS," Proceedings, *Oceans 88,* Vol. 4; R. Stewert and M. Kennedy, "An Analysis of U.S. Tanker and Offshore Petroleum Production Spillage Through 1975," Report to D.O.I. Office of Policy Analysis, Martingale, Inc., Cambridge, MA. 1978.)

spills from domestic vessels. The foreign tanker figure is based on historic spills by foreign vessels in U.S. waters.

When oil is spilled, estimating its likely fate is important because this will help determine the type and extent of damages. For example, spills that come ashore could affect the value of beach use to recreationists, impose losses on tourism businesses, and require costly shoreline clean-up operations. In contrast, spills that remain at sea will not impose beach losses and may not require substantial clean-up efforts.

The estimates of the chance that a spill in a given area reaches shore used in this section draws on the summary results of several hundred to 2,000 simulated oil spill trajectories for hypothetical production and transportation spills for each OCS area for each season (Table 4-1).[12] The number used in effect is the weighted average risk that oil spilled within an area will contact the shoreline, based on the likely production sites, transportation routes, and prevailing winds and currents for each area used in designing the simulations. The results of these simulations are adopted from information presented in recent environmental impact statements for each OCS area, and they indicate, for example, that the chance of spills reaching shore ranges from about 85 percent for the Gulf of Alaska to about 10 percent for North Atlantic OCS spills.

Risk of Nonspill Costs

In addition to oil spills, OCS development may cause several other environmental costs. The presence of substantial production and platform facilities could preclude fishing in a particular area, and, in addition, commercial fishing gear could be damaged by pipelines and bottom obstructions. For example, projected oil and gas facilities development on Georges Bank, one of the country's most productive fishing areas, has been projected to result in a peak annual loss of catch of $114.8 thousand.[13] Regarding gear losses, the National Marine Fisheries Service (NMFS), which administers the special fund established by the OCSLA to compensate fishermen for OCS-related gear damages, reports that in fiscal year 1985 fishermen filed 199 claims for $1.6 million against the fund.[14]

If pipelines are used to transport oil and gas to shore, coastal wetlands could be disrupted through direct impacts (physical conversion) and through indirect impacts (effects on natural processes due to direct effects). Historically, this has been a substantial issue in the Central Gulf of Mexico, although recent evidence suggests that OCS operations are responsible for only 14–16 percent of direct impacts on wetland losses and 5–18 percent of indirect impacts.[15] In addition, air quality issues could arise, particularly in Southern California, which has existing air quality problems and where OCS production could take place relatively close to shore. Finally, OCS development could impose planning and infrastructure costs that could be important to relatively undeveloped areas lacking port, airport, road, school, and other public health and safety facilities.

Nonoil spill costs depend more on the scale of total activity—that is, on the magnitude of both oil and natural gas resources developed—rather than simply on annual OCS oil production. Further, possible costs from lost wetlands will depend importantly on such economic-technological considerations as whether pipelines rather than vessels will be used to transport oil and the number of pipelines used, since pipelines that traverse coastal wetlands can cause direct and indirect loss of wetland areas. A further complication arises since excess capacity at existing facilities (pipelines, support bases, and port facilities) might be sufficient to accommodate the demands from incremental OCS production.

Unit Cost of Oil Spill and Nonspill Costs

Table 4-2 summarizes the unit cost estimates used for each of the costs considered. Two general approaches were used to arrive at these estimates. In many cases, estimates from available case studies were adapted to quantify unit costs, and generally these costs were scaled to each geographic area to reflect differences among the OCS areas. For example, estimates of oil spill clean-up costs for case studies of spills were not available for Alaska; costs to clean up spills in Alaska were assumed to be 45 percent higher than the lower 48, based on an appropriate price index.[16] Case studies of the *Exxon Valdez* and Glacier Bay spill should provide considerable new insight into the cost of clean-up in Alaska that is likely to expand the upper range of clean-up costs for oil that comes ashore.

Table 4-2
Unit Costs

Summary of unit cost estimates used in analysis of costs of proposed final OCS five-year leasing program

Cost category	Cost per indicated unit ($1987)[a]
Oil Spill Costs	
1. Clean-up and control costs	
a. Production platform	
(i) Oil comes ashore	$225–326 per bbl ashore
(ii) Oil remains at sea	$102–149 per bbl spilled
b. Pipeline	
(i) Oil comes ashore	$222–321 per bbl ashore
(ii) Oil remains at sea	$ 62–91 per bbl spilled
c. Tanker	
(i) Oil comes ashore	$228–331 per bbl ashore
(ii) Oil remains at sea	$ 21–32 per bbl spilled
2. Commercial fishing	
(i) Direct losses	neg–$150 per bbl spilled
(ii) Secondary (multiplier)	$89–273 per $100 loss in commercial fishing income
3. Tourism industry and recreation losses	
(i) Recreation losses	$169–675 per bbl spilled reaching shore[b]
(ii) Tourism industry losses	$0.9 per dollar of recreation losses
4. Ecological costs	$41.7–319 per bbl spilled[b]
5. Subsistence losses	$134 per bbl spilled for coastal Alaskan OCS
6. Value of lost oil	$29.00 per bbl spilled[c]
7. Other costs	
a. Legal-administrative costs	$18 per bbl spilled
b. Research costs	$8 per bbl spilled for spills $\geq$ 1,000 bbls
c. Property value losses	$45.5 per bbl spilled reaching shore for "lower 48"
	$5 per bbl spilled reaching shore for Alaska
Nonspill Costs	
1. Commercial fishing	
a. Area preemption	$0.5 million per BBOE produced
b. Gear losses	$1.4 million per BBOE produced
2. Air pollution	$0.0195 per BOE—So. Calif.
	$0.059 per BOE—Cent. Gulf
	$0.0029 per BOE—all other OCS areas
3. Wetlands	$11,611–58,280 per acre lost[b]
4. Infrastructure costs	neg.–$24.0 million per BBOE[b]

[a] See text and U.S. Department of the Interior, *Five Year Leasing Program*, Appendix G, for discussion of the derivation of the individual unit cost estimates.
[b] The indicated range reflects the range of unit cost estimates used for different OCS planning areas.
[c] This is the high starting oil price which is assumed to increase by 1 percent in real terms annually.

As another example, the value of wetlands lost because of OCS operations was estimated for each OCS area using the basic approach of Gupta and Foster.[17] Their approach imputes a value to each of the individual major characteristics or services provided by wetlands: wildlife, visual, cultural, flood control, and water supply. To adapt this approach to OCS issues, coastal wetlands were presumed to provide the following services: wildlife, visual, cultural, flood control, and habitat/nursery services for marine fisheries. Various studies were surveyed to establish costs for losing a unit of each of these characteristics. The use of this characteristics-type approach, leads to a preservation value for wetlands ranging from $11,611 to $58,280 per acre. These estimates are far higher than more detailed site-specific estimates found in the literature, and their use reflects the overstated cost approach used to quantify environmental costs.[18]

In other cases, available data on particular costs of OCS development were assembled to develop a unit cost estimate for particular environmental costs that could be expected to result from OCS operations. For example, experience with claims filed by commercial fishermen in the Gulf of Mexico under the Fishermen Contingency Fund was used to quantify possible gear losses in OCS areas. Generally speaking, in each case the estimates developed were deliberately biased to overstate costs in order to present a conservative, high estimate of social and regional costs.

Selected Results for Social and Regional Costs

Using the methodology and data outlined above, estimated social costs for the $14 per barrel oil price range from $41.8 million for the central Gulf of Mexico to less than $1 million for several OCS areas with minor amounts of resources, as shown in Table 4-3. These costs are stated in present value terms where dollar flows in future years are converted to a present value using a discount rate of 8 percent as is required for federal projects of this type. Note that all values are in constant 1987 dollars except for the starting price of $14, which is in 1984 dollars.

The importance of netting out the environmental costs that would result from spills from foreign tankers in the absence of OCS production is made clear by examining the results in Table 4-3. For example, if the backout of imported oil is ignored, OCS production in the central Gulf of Mexico is estimated to lead to social costs of $48.8 million. However, production in this area reduces imports in this and other OCS areas and thereby avoids environmental costs of $7.0 million, so that the net social cost amounts to $41.8 million.

Estimates of regional costs for the $14 per barrel case are presented in Table 4-4. These are the costs to the residents of each OCS area as a result of OCS production in all areas. Because of the specific regional accounting stance adopted in order to reflect the geographic distribution of costs, some items included in regional costs are not included as social costs. The principal example is tourism business losses. Tourism losses that would result from an oil spill affecting a section of coastline will be offset by increased spending elsewhere; hence losses to tourism businesses at the regional level are assumed to net out from a national viewpoint.[19] Also, compensation payments for gear losses and for some oil spill costs are deducted from regional costs, although these items are transfers from a national perspective.

Regional costs range from $34.4 million for the central Gulf of Mexico to zero and

Table 4-3
Potential Social Cost

Summary of the present discounted value of estimated potential social costs for each OCS planning area: $14 per barrel starting price in millions of 1987 dollars

Area	(1) Oil spill costs	(2) Nonspill costs	(3) = (1) + (2) Gross social costs	(4) Less: cost avoided from reduced imports[a]	(5) = (3) − (4) Total net discounted social costs[b]
CGOM	33.2	15.6	48.8	7.0	41.8
WESTGOM	15.7	18.9	34.6	4.2	30.4
S.CALIF	4.9	2.8	7.7	1.6	6.0
N.CALIF	1.5	2.7	4.2	0.6	3.6
EASTGOM	1.4	1.8	3.2	0.4	2.8
CEN.CALIF	1.3	1.4	2.7	0.5	2.2
S.ATLAN	0.7	1.7	2.4	0.2	2.1
MID-ATLAN	0.5	0.8	1.3	0.1	1.1
N.ATLAN	0.0	0.5	0.5	0.0	0.5
ORE/WASH	0.2	0.2	0.5	0.1	0.4
NAVARIN	0.0	0.0	0.0	0.0	0.0
BEAUFRT	0.0	0.0	0.0	0.0	0.0
CHUKCHI	0.0	0.0	0.0	0.0	0.0
ST.GEORGE	0.0	0.0	0.0	0.0	0.0
N.ALEUTIAN	0.0	0.0	0.0	0.0	0.0
GULF ALASKA	0.0	0.0	0.0	0.0	0.0
NORTON	0.0	0.0	0.0	0.0	0.0
KODIAK	0.0	0.0	0.0	0.0	0.0
HOPE	0.0	0.0	0.0	0.0	0.0
SHUMAGIN	0.0	0.0	0.0	0.0	0.0
COOK INLET	0.0	0.0	0.0	0.0	0.0
STR OF FLOR	0.0	0.0	0.0	0.0	0.0

[a] Estimated potential social costs avoided to the nation as a whole from reduced needs for imported oil.
[b] Zero costs indicated in the table occur because area contains negligible leasable resources and estimated costs are less than $.1 million.

are lower than social costs for some OCS areas but higher for others. For example, consider the central Gulf of Mexico. The compensation paid to residents for gear losses and for oil spill costs, and the environmental cost savings realized when OCS oil from all areas displaces imported oil in this important petroleum refinery area, are sufficiently high that the net regional costs are lower than the net social costs for this area. This is so despite the fact that the oil spill and nonspill costs to the region are greater than their social cost counterparts because they include such purely regional costs as tourism business losses.

On the other hand, for some OCS areas, regional cost is greater than social cost. For example, the Southern California area is expected to have higher regional costs than the associated social costs of production. This is due to the relatively large oil spill costs resulting from transportation of crude oil from all Pacific areas. Regional costs resulting from OCS production outside Southern California are not social costs of de-

Table 4-4
Regional Costs

Estimated potential regional costs by area from development of all areas: $14 per barrel starting price in millions of 1987 dollars

Area	(1) Oil spill costs in region	(2) Nonspill costs in region	(3) Compensation to region	(4) Import costs backed out of region	(1) + (2) − (3) − (4) Potential net costs to the regions[a]
CGOM	44.6	15.6	16.2	9.6	34.4
WESTGOM	22.1	18.9	8.0	8.4	24.5
S.CALIF	7.2	2.8	2.3	0.4	7.3
N.CALIF	2.0	2.7	0.7	0.0	3.9
EASTGOM	1.9	1.8	0.7	0.0	3.0
CEN.CALIF	2.4	1.4	0.6	0.1	3.0
S.ATLAN	0.7	1.7	0.4	0.1	1.9
ORE/WASH	0.6	0.2	0.1	0.2	0.5
N.ATLAN	0.0	0.5	0.0	0.1	0.4
MID-ATLAN	0.5	0.8	0.3	1.0	0.1
NAVARIN	0.0	0.0	0.0	0.0	0.0
BEAUFRT	0.0	0.0	0.0	0.0	0.0
CHUKCHI	0.0	0.0	0.0	0.0	0.0
ST.GEORGE	0.0	0.0	0.0	0.0	0.0
N.ALEUTIAN	0.0	0.0	0.0	0.0	0.0
GULF ALASKA	0.0	0.0	0.0	0.0	0.0
NORTON	0.0	0.0	0.0	0.0	0.0
KODIAK	0.0	0.0	0.0	0.0	0.0
HOPE	0.0	0.0	0.0	0.0	0.0
SHUMAGIN	0.0	0.0	0.0	0.0	0.0
COOK INLET	0.0	0.0	0.0	0.0	0.0
STR OF FLOR	0.0	0.0	0.0	0.0	0.0

[a] Zero costs indicated in the table occur because area contains negligible leasable resources and estimated costs are less than $.1 million.

veloping oil in Southern California but they are costs to residents of the area. Important regional costs include tourism business losses from oil spills; these losses are reduced somewhat by compensation payments for business losses and by reduced imports.

Sensitivity Analyses

Recognizing the difficulties and uncertainties inherent in measuring social costs, two types of sensitivity analyses were done in addition to considering different prices of oil. The first examines how the social and regional results would differ if some of the unit costs proved to be even larger than the overstated cost figures used to derive the base case results. For this purpose, commercial fishing losses, wetland losses, and ecological effects were allowed to be up to 50 percent higher than the unit cost estimates presented in Table 4-3. These costs were selected both because they are quantitatively important for most areas and because their measurement is subject to considerable uncertainty.

The results of the extreme case sensitivity analysis in which each of the three costs is presumed to be 50 percent higher leads to a less than 30 percent increase in total net social costs. For example, for the central Gulf of Mexico for the resource estimate at $14 per barrel, social costs increase from $41.8 million to $50.2 million. The reason why net social costs increase by less than 30 percent is that increasing the unit costs for commercial fishing and ecological costs from OCS oil spills also increases the social cost savings from replacing imported oil.

The second set of sensitivity analyses assesses how sensitive the base case results are to the choice of a discount rate. The results described in the above paragraphs are based on the assumption that future costs are to be converted to a present value using an administratively determined rate of 8 percent. One could argue that a lower rate would provide a more accurate measure of the real social discount rate. For example, in a recent paper summarizing the results of an environmental study of damages from PCB pollution in the bottom sediments of New Bedford Harbor, a 3 percent real discount rate was used.[20]

If a 3 percent discount rate is used with the high resource case at $29 per barrel, the social costs for each area increase substantially (Table 4-5). For example, for the central Gulf of Mexico, total discounted social costs more than double from $42.3 million to $85.1 million for the high resource case. For the western Gulf of Mexico, the area with the second highest social costs, the discounted social costs increase from $35.4 million to $62.8 million when the discount rate is lowered from 8 to 3 percent.[21]

Regional costs follow a similar pattern. For example, for the central Gulf of Mexico (the OCS area with the highest cost), regional costs increase from $27.7 million to $54.4 million when the discount rate is lowered from 8 percent to 3 percent.

AN INTEGRATED INTERDISCIPLINARY MODEL

This analysis explores how the estimates of social cost might vary if a different approach than the simplified, unit cost approach was used to estimate environmental costs. For issues as complex and difficult as determining environmental costs, this type of alternative analysis is especially important because it is recognized that the unit cost approach suffers from several shortcomings.

Use of a unit cost approach is based on highly simplifying assumptions regarding transferability of case study results from one area to other areas and situations. Further, in some cases the original estimate may suffer from its own conceptual and empirical shortcomings. This is particularly true for such costs as commercial fishery losses from oil spills and ecological costs, which are extremely difficult to quantify. Moreover, the use of unit cost estimates implicitly assumes that there are no thresholds and that costs increase in a linear relation with the amount of oil spilled. In fact, very small spills may result in negligible or very minor damages, and damages could increase at an increasing or decreasing rate when greater quantities are spilled. Use of an overstated cost approach to some extent compensates for the need for precise estimates, provided, of course, that the unit cost estimates used, together with the other assumptions that

Table 4-5
Sensitivity Results: 3% Discount Rate

Summary of the present discounted value of estimated potential social costs for each OCS planning area: $29 per barrel starting price—3% discount rate in millions of 1987 dollars

Area	(1) Oil spill costs	(2) Nonspill costs	(3) = (1) + (2) Gross social costs	(4) Less: cost avoided from reduced imports[a]	(5) = (3) − (4) Total net discounted social costs[b]
CGOM	72.7	30.1	102.8	17.63	85.14
WESTGOM	43.5	32.7	76.2	13.33	62.82
S.CALIF	20.7	9.5	30.2	7.87	22.33
S.ATLAN	3.9	6.0	9.9	1.65	8.23
NAVARIN	20.3	12.9	33.2	6.94	26.29
EASTGOM	8.5	4.5	13.1	2.49	10.59
BEAUFRT	10.1	2.2	12.3	3.24	9.03
CHUKCHI	11.2	0.7	11.9	3.94	7.95
CEN.CALIF	4.7	3.0	7.7	2.10	5.59
N.CALIF	6.2	6.0	12.2	2.74	9.42
ST.GEORGE	2.7	5.0	7.7	1.15	6.52
MID-ATLAN	2.0	2.0	3.9	0.63	3.32
N.ATLAN	0.3	1.1	1.4	0.16	1.19
ORE/WASH	0.5	0.4	0.9	0.19	0.72
N.ALEUTIAN	0.4	0.3	0.7	0.10	0.60
GULF ALASKA	0.4	0.4	0.9	0.09	0.79
NORTON	0.5	0.5	1.0	0.15	0.81
KODIAK	0.0	0.0	0.0	0.00	0.00
HOPE	0.0	0.0	0.0	0.00	0.00
SHUMAGIN	0.0	0.0	0.0	0.00	0.00
COOK INLET	0.0	0.0	0.0	0.00	0.00
STR OF FLOR	0.1	0.5	0.6	0.05	0.54

[a] Estimated potential social costs avoided to the nation as a whole from reduced needs for imported oil.
[b] Zero costs indicated in the table occur because area contains negligible leasable resources and estimated costs are less than $.1 million.

are deliberately set to overstate costs in fact portray an overstated cost result. Nonetheless, it is important to examine how environmental costs might vary if a conceptually more sophisticated approach is used.

For this purpose, the recently published Natural Resource Damage Assessment Model for Coastal and Marine Environments (NRDAM/CME) was used to measure specific categories of damages from OCS oil spills. The NRDAM/CME was developed for simplified natural resource damage assessments under the Comprehensive Environmental Response, Compensation and Liability Act of 1980 (hereafter referred to as CERCLA). This new approach, outlined below, provides a conceptually superior framework for estimating natural resource losses based on the concept of a damage function. Its use tests the reasonableness of the assertion that the simplified unit cost approach provides an overestimate of environmental costs.[22]

The NRDAM/CME is an integrated, interdisciplinary model that simulates the diffusion and degradation of oil spilled in the marine environment, some of its short- and long-term biological effects, and the resulting loss in the *in situ* use value—value in place—of specific categories of natural resources (Fig. 4-3). Natural resources losses are measured for: (1) commercial and recreational fisheries, (2) waterfowl, sea and shore birds, (3) fur seals, (4) lower trophic biota, and (5) public beaches.

The NRDAM/CME model is not ideal; large spills are highly complex, situation-specific events, and the model itself involves many simplifications. For example, biological data for each season are averaged over an area within one of the many environment-type categories the user specifies. Nonetheless, given that projected OCS activity from a proposed leasing schedule will take place over many years and that the analysis is being done on a planning area basis, it is not possible to strive for precision in any event. Further, the NRDAM/CME offers important advantages as a tool for assessing environmental risks from oil spills in that (1) it is conceptually superior to the unit cost approach (2) it simulates damages (to fisheries, for example), which rarely,

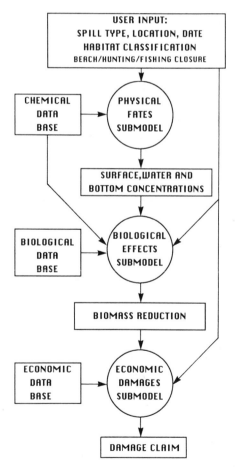

Figure 4-3. Overview of Type A natural resources damage assessment model for coastal and marine environments.

if ever, can be observed following a spill, and (3) it provides a consistent framework for assessing potential damages across OCS planning areas.

To use the NRDAM/CME, it is necessary to specify the type of oil spilled and the amount, the season, and location of the spill, and other site-specific parameters such as water depth, air temperature, distance from shore or other boundaries of concern, and wind and current direction and speed. With this information, the physical fates submodel simulates the spreading and degradation of the oil on the surface as a slick and in the upper and lower water column and in the bottom sediments, if appropriate. Concentrations of the pollutant in the water column and bottom sediments are calculated over time; this information—together with the surface area of the slick—is then fed to the biological effects submodel.

The biological effects submodel quantifies specific short- and long-term effects. Short-term effects include acute mortality to adult fish, shellfish, birds, and fur seals. Long-term effects include lost biomass because adults are killed by the spill, and lost recruitment to fisheries caused by the death of juveniles and larvae. In addition, a simple food web model is used to quantify losses to fish, birds, and fur seals due to loss of their potential food supply.

To determine the biological resources at risk from a spill, a substantial data base was assembled. It is comprised of nine fish and shellfish categories, which include virtually all commercially and recreationally important species grouped by trophic level and habitat. Also included in the data base are separate categories for waterfowl, shore and sea birds, and fur seals (the only mammal for which use value data could be assembled). Number of larvae per unit area by species category and measures of primary and secondary productivity were also obtained. These data were assembled for ten geographic provinces. Within each province a distinction is made between subtidal and shoreline spills, open ocean vs. estuary spills, bottom type, and season. The data base contains distinct biological data for a total of 364 province-environment-season types.

To apply this model to estimate OCS environmental costs, each of the OCS regions was assigned to one of the larger geographically corresponding NRDAM/CME provinces. Spill locations were set at distances to shore corresponding to the general area of petroleum potential for each area. To assess shoreline effects, the coastline of each OCS area was categorized by the predominant shoreline type for the area as developed in U.S. Department of the Interior studies.[23]

Oil spilled in Atlantic and Gulf of Mexico OCS areas was presumed to be medium crude; oil spilled in the Pacific and in the Alaskan OCS areas was presumed to be a heavy crude. The size distribution of spills was determined by using the spill probability distribution for each source (platforms, vessels, and pipelines) relevant to each OCS area. Expected damages were then calculated by combining estimates of damages as a function of spill size (which were generated by the NRDAM/CME model) with expected numbers of spills of various sizes.

The results of this sensitivity analysis presented in Table 4-6 indicate that the estimated environmental cost per barrel of oil spilled using the NRDAM/CME is considerably lower for 18 of the 22 OCS planning areas studied. Hence, for these OCS areas the estimates of social and regional costs described earlier would be lower if the NRDAM/CME results had been used in place of the unit cost estimates for the costs concerned. On the other hand, the NRDAM/CME approach results in somewhat higher costs per barrel spilled for four OCS planning areas, central and northern California and the St. George and Norton basins on the Alaskan OCS. In summary, the results suggest that,

Table 4-6
Comparison of Cost Models

Comparison of the cost per barrel spilled using the unit cost estimates and those obtained using the CERCLA NRDAM/CME

CERCLA province	OCS planning area	Cost per barrel spilled:	
		Unit cost[a]	NRDAM/CME[b]
Acadian	North Atlantic	$282.5	$ 26.7
Virginian	Mid-Atlantic	261.8	13.8
Carolinian	South Atlantic	216.6	4.4
West Indian	Straits of Florida	203.9	60.7
Louisianian	Eastern Gulf of Mexico	127.8	37.6
	Central Gulf of Mexico	330.7	37.1
	Western Gulf of Mexico	172.8	11.4
Californian	Southern California	167.2	27.2
	Central California	173.3	210.4
	Northern California	176.3	210.4
Columbian	Washington and Oregon	257.3	140.2
Fjord	Gulf of Alaska	224.3	8.2
	Cook Inlet	228.7	84.1
	Kodiak	233.0	82.8
	Shumagin	216.4	84.1
Arctic	St. George Basin	233.7	250.7
	North Aleutian Basin	344.2	249.1
	Navarin Basin	49.7	8.6
	Norton Basin	227.4	249.1
	Hope Basin	319.7	32.0
	Chukchi Sea	125.9	30.2
	Beaufort Sea	202.0	30.4

[a] Each number in this column is the sum of the direct commercial fishing loss and the ecological cost per barrel of oil spilled.
[b] The number in this column indicates the weighted average of the expected damages per barrel for each potential oil spill source (platform-tankers-pipelines) for each of the costs considered in the NRDAM/CME except for damages to public beaches. The expected damage per barrel spilled for each source was weighted by the expected number of spills per billion barrels of oil produced or handled and the relative share of oil handled by tankers and pipelines for each area.

compared to the NRDAM/CME model, the unit-cost approach used to measure social and regional costs provides a considerably overstated estimate of environmental costs for the categories considered in over 80 percent of the cases, whereas the environmental costs are somewhat understated compared to the NRDAM model for four areas. Given the stronger conceptual foundation of the NRDAM/CME as compared to the unit cost approach, these results should be integrated into the policy process to assess potential social costs in future policy analysis. At the same time, efforts are needed to refine and improve the NRDAM/CME.

LIABILITY AS A POLLUTION CONTROL INCENTIVE

An important role played by economic analysis in environmental policy is in helping to assess the effectiveness of approaches for controlling pollution on the OCS. Under the OCSLA (Sec. 21), oil and gas operators in federal waters are required to use the best available and safest technology when economically feasible. In addition, firms responsible for offshore spills are held strictly liable for the damages from OCS-related oil spills (Sec. 303).

Whereas the purpose of strict liability established under the OCSLA is to compensate damaged parties, this liability can be viewed as a tax on the external costs covered by the Act, which should cause firms to internalize the expected environmental cost of developing OCS reserves.[24] Thus the OCSLA provides oil companies with clear economic incentives to control for environmental risk through investing in safe technology, altering production rates, and avoiding environmentally sensitive areas.[25] The financial incentives approach provides motivation for company managers to improve safety. By penalizing only those companies that have accidents, strict liability focuses penalties on companies with poor safety records. However, most public debate focuses on imposing detailed regulations and pays little attention to the role that economic incentives can play in controlling pollution.

This section examines the effectiveness of strict liability for oil spill damages in controlling oil spill pollution events. The basic argument is that firms can be expected to perceive their potential liability for environmental damages for oil spills and to respond to that risk. If this argument is correct, then the cash bonus bids that firms offer for tracts in environmentally sensitive areas should be lower than bids in less sensitive areas, after correcting for other factors such as differing resource estimates or development costs. The effect is to transfer money back to a company to compensate it for spill avoidance costs and the likely cost of paying damages.

To test this proposition, the December 1979 North Atlantic (Georges Bank) lease sale #42 was used as a case study. This OCS lease sale is a particularly appropriate application because of the importance of Georges Bank commercial fisheries, the proximity of the sale area to intensively used coastal beaches, and the extent to which the area has been studied. The highly contentious nature of the environmental concerns for Georges Bank is underscored by several delays in the lease sale date, by the litigation preceding the sale, and by the deletion of 28 tracts from the sale—all for environmental reasons. Further, the fact that this was the first sale of leases in the Georges Bank area avoids the potential difficulty of informational asymmetries, whereby the bidding behaviors of firms could be influenced by information acquired at a prior sale.[26]

At the Georges Bank lease sale #42, a total of 31 companies submitted sealed bids for 73 tracts out of the 116 offered. The high bids for individual tracts ranged from $201,000 to $80 million; a total of $817 million in high bids was accepted by the Department of the Interior.

The size of a winning bid for a tract depends upon the company's estimate of the value of the resources believed to be contained in the tract, the number of competing bids, and other factors.[27] If companies do indeed consider their liability for the damages from possible oil spills when assessing the value of a tract, then company bids should also reflect the risk associated with the tract. Bids should be systematically lower in areas of greater potential environmental damage, since firms could be held liable for damages should a spill occur.

To isolate the influence of environmental risk on a company's assessment of the value of a tract, a bidding model was developed and applied to data for the Georges Bank lease sale.[28] The model explains the high bids for each tract as a function of the number of bidders for the tract and the estimated prelease sale value of the tract, as reflected by the U.S. Geological Survey's presale net value estimate. Another variable was included to measure scale economies and information externalities that could be realized by a firm that develops adjacent tracts. Most important, from the perspective of environmental issues, a variable measuring environmental uncertainty (described below) was constructed using the results of the Department of Interior's Environmental Impact Statement (EIS) for the lease sale.

The environmental sensitivity variable measures the potential threat to major resources in the area. In terms of fisheries, the EIS considered the potential threat to cod and haddock spawning areas and to shellfish grounds on Georges Bank. In 1980 the value of landings from these three fisheries amounted to $77 million, about two-thirds of all of the value of landings from Georges Bank. The variable also includes the threat to nearby shorelines. This category of risks is particularly important because of possible compensable losses that could be incurred by coastal residents and by owners and operators of coastal businesses as the result of a spill. In addition, the cost of cleaning up spilled oil that comes ashore can represent one of the largest monetary costs of a spill.[29]

The specific measure of environmental risk is the probability that the trajectory of a spill would strike fishing areas mentioned above or would come ashore. These probabilities are given in the EIS for 11 hypothetical, geographically dispersed oil spill launch sites throughout the Georges Bank lease area, and six of these sites are within the area sold. The variable measuring environmental sensitivity is equal to the sum of the probabilities that a spill would come ashore, strike shellfish areas, and strike cod and haddock spawning grounds, scaled by a factor of 100. This measure ranges from 15 for the least environmentally sensitive tract to 34 for the most sensitive. That is, the sum of the probabilities that a spill that occurs will strike one or more of the designated high-risk environmental areas varies from .15 to .34.

The results indicate that high bids on individual tracts are between $1.67 million (with one bidder) and $4.33 million (with seven bidders) lower for the most environmentally sensitive tracts as compared with the least sensitive tracts, all else being equal. In the aggregate, if all tracts had zero environmental risk, total high bids are estimated to have been $236.7 million higher than what they were. This figure represents a substantially greater response by firms than had been expected. To place this figure in perspective, the expected environmental damages viewed as potentially compensable under the OCSLA from developing the Georges Bank lease area were estimated by the Interior Department to be $101 million.

Several factors may explain this apparent discrepancy. First, damage estimates by the Interior Department did not include oil spill removal costs, nor did they allow for the monetized estimates of damages to noncommercial natural resources, both of which are potentially compensable under the OCSLA. Second, part of the differential may be attributable to company concerns about potential legal actions delaying development in environmentally sensitive areas, or the difference may reflect company concerns about adverse publicity that would inevitably accompany a large oil spill on Georges Bank (particularly if it came ashore). Third, some of the $236.7 million reduction in high bids may reflect company's costs of spill avoidance. Finally, companies may be averse to risk that could be exacerbated by the regulatory uncertainty resulting from the lack of experience with the determination of damages under the OCSLA.

In summary, the test of firms' perceptions of environmental risk as reflected in their actual as compared with their predicted bids for OCS tracts suggests that offshore oil companies do take environmental risk into account. Hence, strict liability may be an important way to encourage firms to avoid pollution through adoption of a variety of possible actions. Since the time of the 1979 Georges Bank lease sale, the NRDAM/CME described earlier has been developed by the Department of the Interior as a simplified approach for assessing polluters for damages under CERCLA. A more extensive, site-specific study of damages could be carried out if the simplified approach provided by the NRDAM/CME is judged not to be appropriate under the criteria of the act. The existence of these new approaches under CERCLA and the procedures recently put into place by federal rules to implement these alternative damage assessment methodologies greatly facilitates the assessment of damages from OCS-related oil spills and should increase the effectiveness of economic incentives in controlling oil spills from OCS-related operations.[30]

The economics perspective provides several key insights into OCS environmental policy issues. First, an approach was presented to help quantify the potential environmental costs that could result from the implementation of a new OCS leasing program. The framework integrated available information concerning area-specific OCS resource estimates, annual production, transport modes and routes, spill rates and trajectories, and unit oil spill and nonspill costs to provide a perspective on the environmental costs of leasing for each OCS area.

The measure of social costs for each area provides information useful for assessing the potential net benefits to the nation from developing each OCS area. The social cost results indicate the trade-offs that the country as a whole faces from OCS development in a given area, unless mitigating measures or other regulatory actions at the federal and state levels are effective in reducing these potential costs. Similarly, the estimate of regional costs suggests the magnitude of costs that could be realized by the residents of each OCS area. Regional costs, in effect, capture the geographic distribution of costs and, thus, address the equity concerns reflected in Section 18 of the OCSLA. This information can assist in making decisions regarding the balancing of benefits and costs among regions.

Throughout this chapter the many difficulties and uncertainties surrounding the measurement of social and regional costs have been emphasized. The use of overstated cost estimates and assumptions that consistently result in higher cost estimates was consciously adopted to reflect these difficulties and uncertainties. In addition, several sensitivity analyses were presented to show how the estimates of social and regional costs would change if alternative costs even higher than the overstated costs were employed, or if a lower discount rate was used. A particularly important extension was the use of the recently developed Natural Resource Damage Assessment Model for Coastal and Marine Environments (NRDAM/CME). In spite of the simplifications necessarily associated with the use of such a model, this approach represents a major advance in the modeling of the environmental costs of marine pollution incidents as compared with a unit cost approach. The model simulates the spreading and degradation of an oil spill, quantifies some of the resulting short- and long-term biological injuries, and provides a measure of the associated economic damages.

In regard to liability rules, economics reminds us that firms change behavior when faced with higher costs. The liability rules contained in OCSLA appear to have reduced government auction income to compensate firms for incurring spill avoidance costs and for the possibility of paying damages and litigation associated with OCS exploration and production.

Finally, an unavoidable difficulty that arises when attempting to provide any assessment of the environmental risk from future OCS oil and gas is that political, economic, and technological conditions can change substantially over time. Several recent developments illustrate this point. For example, in response to concerns about air quality in Southern California, the Department of the Interior has proposed very stringent restrictions on the release of air pollutants from OCS facilities in this OCS area.[31] Also, the EPA has proposed restrictions on the discharge of drilling muds and cuttings, and regulations for produced waters are being developed.[32] Finally, the 240,000-barrel oil spill from the *Exxon Valdez* and several other large and highly publicized oil spills in the lower 48 (none resulting from OCS oil) has created considerable momentum in the United States Congress to improve tanker safety practices and clean-up operations and to strengthen and extend liability for clean-up costs and environmental damages. To the extent these regulations are implemented and effective, the risks from OCS development will be reduced and additional compensation for externally imposed costs will be available.

Chapter 5
Fair Market Value

Congress has specified the goal of receiving fair market value for leases. Exactly what fair market value means remains a debatable issue both for offshore and for onshore leasing. Regardless of the definition used, however, fair market value determines the allocation of income between the government and the private sector. This chapter looks at the definition of fair market value accepted by the court and discusses several policy analyses regarding the receipt of fair market value.

The role of policy analysis in assessing fair market value includes the bid adequacy criteria that determine whether a bid is accepted or rejected, assessing rates of return from the purchase of leases, and the regression analysis of lease sales. In this chapter the analysis of baseline rates of return prior to the major change in the rate of leasing in 1983 is followed by an analysis of a charge that the 1983 change in policy led to a $7-billion loss in bonus revenue to the government. Finally, a prospective policy proposal that would affect the receipt of fair market value is presented. This proposal suggests that the auction process be open to those wishing to delay leases and quantifies the possible cost to interest groups of purchasing such delays.

BID ADEQUACY CRITERIA

In *California vs. Watt (II)* the federal district court ruled on the principle and on the means of implementation used by the Department of the Interior in achieving fair market value. The court accepted Interior's definition that

> Fair market value is defined as the amount in cash, or in terms reasonably equivalent to cash, for which in all probability the property would be sold by a knowledgeable owner willing but not obligated to sell to a knowledgeable purchaser who desired but is not obligated to buy.[1]

This is a standard definition; the controversy lies in how it is implemented.[2] Prior to the change in the pace of leasing in 1983 (often referred to as the change from tract selection to areawide leasing procedures), the Department of the Interior conducted a present value evaluation of every tract offered for lease. In general, if the bonus bid for the tract exceeded the mean value of Interior's evaluation, then fair market value was received. When Interior increased the acreage offered by a factor of 15 in 1983, it concluded that it was impractical to evaluate every tract *prior* to the lease sale. Instead, it would evaluate selected tracts actually bid on after the sale. These rules for accepting or rejecting the bids (which have not always been bonus bids since Congress mandated experimentation with a variety of leasing systems) are known as bid adequacy rules.

The particular rules that the court reviewed in *Watt (II)* are not those currently in place (though the court indicated that a variety of ways might reasonably implement the goal of achieving fair market value). Particularly important was the court's observation that fair market value did not mean the maximization of revenue to the government. The Department of the Interior could rely jointly on a competitive bidding process and a supplemental evaluation procedure—the bid adequacy criteria. In particular, a decline in the number of bids or even a decline in the amount of a bid did not necessarily indicate failure to receive fair market value as past revenues might have exceeded fair market value.[3]

The current bid adequacy procedures continue to rely on the congressionally mandated sealed bid auction as a competitive process. The Minerals Management Service has returned to auctions defined by a minimum bid, a fixed royalty rate, and a lump sum bonus after studying congressionally mandated experiments with alternative systems from 1978 to 1983.[4] Current procedures require that all bids must be submitted by a qualified bidder and exceed the minimum bid per acre. The minimum bid was, with few exceptions for frontier areas, increased from $25 to $150 per acre in 1982 and then reduced to $25 per acre in 1987.

The remainder of the bid adequacy rules contains a variety of provisions that depend on the type of tract and the degree of competition as measured by the number of bidders. First, consider that there are two major categories of tracts: the first category includes wildcat and proven tracts (proven tracts have been drilled but relinquished to the government), which are considered to be the most uncertain for current exploration and development. The second category includes drainage and development tracts that are geologically associated with an existing discovery or production. For wildcat and proven tracts, the high bid is automatically accepted as being indicative of fair market value if there are three or more bids or if the tract is judged to contain nonviable quantities of oil and gas—an important application of the analysis of resources. In order to estimate their expected present value, remaining wildcat and proven tracts with fewer than three bids are evaluated based on Department of Interior calculations.

All drainage and development tracts receive a present value evaluation. One justification for evaluating all drainage and development tracts is that the company adjacent to the tract has more information than other bidders. This asymmetry of information affects the likelihood of other firms bidding and the expected value of their bids.[5]

For tracts receiving a present value evaluation, there is a hierarchy of tests to determine if the bid represents fair market value. The first test is whether the bid exceeds the expected present value. (The Department of Interior's term for this is the mean range of value.) If so, the bid is accepted. If the bid does not exceed the expected present value, a second test is conducted where the bid is compared with the expected value if it is sold one year in the future (this is called the delayed mean range of value). If the bid exceeds this expected, delayed present value, it is accepted. This last test is based on the belief that knowledgeable sellers would consider that they have to wait at least one year for the opportunity to lease the tract again during which time some interest earnings are lost (even though prices may also change during that year). Finally, if the bid does not exceed the delayed expected present value, but other bids were received on the tract, then a third test constructs a geometric average of the government evaluation, and of the lower bids, on the presumption that the lower industry bids also contain knowledge about the uncertain value of the tract. If the high bid exceeds this geometric average, it is accepted. If a bid fails to exceed any of these criteria, it is rejected as not achieving fair market value. From 1983 to April of 1986, a total of 3,531 tracts received bids of which 49 percent received evaluations by the Department of the Interior. Of the total number of tracts bid on, 7 percent (253) were rejected as not exceeding the bid adequacy criteria.[6]

The bid adequacy procedures are an important process in demonstrating achievement of fair market value. Their implementation requires a quantitative analysis to compute the several measures of expected value. The Minerals Management Service has developed a simulation program called MONTCAR (for Monte Carlo), which incorporates tract-specific geologic information, an assumed exploration and development scenario, forecasts of prices, and the time span of future production. Many of the same issues used in the estimation of undiscovered resources apply to the estimation of the present value of a tract. In addition, engineering and economic uncertainty is added onto the geologic uncertainty associated with the estimation of resources.

As with resource estimation, there is continuing controversy over the validity of the MONTCAR model. Companies whose bids are rejected on the basis of the model can appeal the rejection. One validation study of the predictions of a model similar to MONTCAR was conducted by Edward Erickson during an appeal of rejections for onshore leasing in Alaska by Exxon Corporation. In a hearing before the Department of the Interior Board of Land Appeals, evidence was presented of the poor correlation between estimates from the government's valuation model and those submitted by industry.[7] Whereas many questions about model validation remain unanswered, MONTCAR is an example of a quantitative model that significantly affects the implementation of policy.

At a more abstract level, auctions are studied extensively in economic theory.[8] Much of the theoretical and empirical work focuses on a single object auction where the value of the object is unknown, but if known, the value would be the same to any buyer. The bid adequacy rules are consistent with the theoretical result of the allocation of value in these auctions: the more bidders the larger the proportion of value captured by the seller; as a result, the price paid converges to the true value of the object.[9] However, there is as yet no consensus on the correct model for OCS leasing, which in fact involves multiple-object auctions, often with asymmetric information, as well as sequential lease sales.

Empirical studies of auctions are limited by the lack of information on the value estimate of the bidder. A recent breakthrough in predicting all bids on the OCS (not just the high bid) as a function of information has been negated by the change in bid adequacy rules that limited the data available.[10] Both the theoretical and the empirical study of bidding and its effect on income received by the seller remain topics on the frontier of research.

In a broader context strong public sentiments exist regarding fair market value. That sentiment seems to imply that large companies should receive as little income as possible while still exploring and producing and that the government should receive the remainder of the income. In economic terms, a statement that a company should earn just the amount necessary to invest in oil and gas production and nothing more is called a normal rate of profit.

From a variety of studies, Walter Mead and coauthors have studied the rate of return earned by oil and gas companies in their OCS operations.[11] Mead and his colleagues studied all OCS leases from 1954 to 1969—necessarily stopping a number of years ago so that a history of production data would be available for each tract. By utilizing the production history, they not only calculated revenues for the historical period, but also determined more accurately the further production that might occur until abandonment. The authors estimated costs for each tract and projected prices until the time of abandonment as well as conducted sensitivity studies of the effect of price changes. They concluded that oil and gas companies, in fact, earned a slightly lower after-tax rate of return on OCS leases, 10.74 percent, than the rate of return earned by all manufacturing firms, which was 11.7 percent in the period from 1954 to 1983. From this aggregate analysis of all tracts leased, the authors concluded that "The close correspondence between after-tax rates of return for OCS lessees and for manufacturing industries generally implies that lessees have not been able to profit at the expense of the federal government by acquiring offshore leases at bargain prices. . . . The auction bidding process and level of competition produced outcomes consistent with the government's objective of receiving 'fair market value'."[12]

Mead and coauthors do report some differences in the mean return depending on the number of bidders and the type of lease. In particular, the lowest after-tax rate of return, 9.32 percent, was earned on leases with five or more bidders, whereas bidders on drainage leases, where asymmetric information is most likely to occur, earned after tax returns of 14.59 percent.

AREAWIDE LEASING AND FAIR MARKET VALUE

The work by Mead and coauthors[13] was prior to the major policy changes of 1983—the change to areawide leasing. This change in policy was expected to lead to reduced competition on tracts, and hence possibly to lower bids. Members of Congress and many interest groups have been concerned about the possible loss of revenue to the government from the change to areawide leasing. The most visible critique has been a General Accounting Office study, which concluded that the change to areawide leasing had reduced bonus bids by $7 billion in only three years.[14] The Department of Interior responded by noting that a possible decline in bonus bids was anticipated but that the majority of the decline was probably due to changes in economic conditions as prices trended downward from 1983 until 1986 when they took a precipitous fall and then recovered in 1987. The Department also emphasized the importance of total revenues received, which includes royalties and taxes, and, even more importantly, the timing of those revenues.[15] In the five-year plan, the Department of the Interior conducted an analysis of the break-even point necessary to indicate a decline in total government revenues from areawide leasing.[16] Interior concluded that even if the declines in the bonus bids as estimated by the General Accounting Office were correct, these declines would have been insufficient to offset the increased revenues associated with the earlier receipt of taxes and royalties associated with offshore leasing.

Whereas it is true that in the court's eyes, a decline in bonus bids is not proof of failure to receive fair market value and that if one wishes to evaluate revenue losses one should consider all sources as well as their timing, it is useful to analyze the results of the General Accounting Office. The discussion can highlight how the first numbers released to the press often receive the most coverage and stay fixed in people's minds, and also how technical improvements in a policy analysis can lead to very different conclusions.

Did Areawide Leasing Decrease Bonus Bids?

The estimate reported by the General Accounting Office of a $7-billion loss in bonus bids was based on a relatively standard regression analysis. The General Accounting Office used data to estimate an average, in this case the average change in the high bid for each tract as a result of areawide leasing. What is particularly valuable about regression analysis is that it estimates an average (and a measure of dispersion—the variance) after taking into account the change in other variables that might have affected an average computed in a simpler way. The regression analysis is said to estimate the conditional average, the average conditional on the value of other variables. This is particularly important in offshore leasing during these years because of changes in prices, interest rates, and the value of the tracts being leased. The result of the regression analysis is one or several equations that quantitatively relate the data of interest, particularly the high bid, to these other variables, including a measure of the change of policy.

In order to implement the regression analysis, the General Accounting Office (GAO) had to make some assumptions about the auction process. That process generates the observed data of the number of bids and the high bid, as well as providing a plausible mechanism to link the change in policy to bidding behavior.

The basic specification chosen by the GAO was that the change to areawide leasing shifted the constant term in an equation for the number of bidders, whereas each of two equations (the number of bidders and the high bid) included variables that were primarily "selected to control for differences in the quality and value of the tract being offered for lease and for differences in the amount or cost of funds used for bidding."[17] These control variables were measures of: (1) the estimated value of the tract generated by the government, (2) the number of bids per tract, (3) a dummy variable (one that takes on two values) for general location, (4) three dummy variables for different lease terms (i.e., different royalty terms), (5) six dummy variables for calendar years, (6) a variable for the proportion of joint bids (several firms jointly submitting one bid) on a given tract, and (7) the intercept shift term for the change in policy.

Conceptually, the GAO estimation procedure can be illustrated by a system of two equations with two unknowns, Y_1 and Y_2, and two known variables, X_1 and X_2. This structural system, analogous to but simpler than the structural system for the number of bids and the high bid equation, is:

$$Y_1 = a_0 + a_1 Y_2 + a_2 X_1 \tag{1}$$
$$Y_2 = b_0 + b_1 Y_1 + b_2 X_2 \tag{2}$$

The GAO argued that b_1 equaled zero and estimated each equation separately using ordinary least-squares.[18] If, for instance, X_2 is a dummy variable that shifts the intercept during areawide leasing, then b_2 is a measure of the initial policy effect on Y_2. This initial effect can be substituted (as a change in Y_2) into equation 1 to obtain the total effect of the policy change. This is, in fact, the method used by the General Accounting Office to obtain their single estimate of the $7-billion loss.

An alternative estimation strategy is to substitute equation 2 directly into equation 1 prior to estimation. Solving the equations yields the reduced form for Y_1 in terms of the known variables, X_1 and X_2 (a similar result holds for Y_2).

$$Y_1 = \left(\frac{a_0 + a_1 b_0}{1 - a_1 b_1}\right) + \left(\frac{a_1 b_2}{1 - a_1 b_1}\right)X_2 + \left(\frac{a_2}{1 - a_1 b_1}\right)X_1 \tag{3}$$

The coefficient for each X in equation 3 (one of which might be a measure of areawide leasing) now includes the *total* effect of each variable on Y_1 where the total effect incorporates the impact on Y_2 as well.[19] Importantly, the total effect can be tested directly using standard hypothesis tests on individual coefficients. In contrast, the procedure used by the General Accounting Office made hypothesis testing of the total effect more difficult and, in fact, was not conducted by the GAO.

To illustrate the importance of these technical differences in a policy analysis, the same type of data used by the General Accounting Office was used to estimate the reduced form model of equation 3.[20] The specification closely follows, but is not equivalent to, the reduced form specification of the GAO model. Key explanatory variables are the same: the policy variable, areawide, is modeled as an intercept shift term; the AAA bond rate is used as the interest rate, and the government's measure of value is used. A slight modification of the GAO model uses both the current level of oil prices and a measure of expected price change (which turns out to be statistically insignificant). Calendar year dummy variables that shift the intercept for each year were a major variation of the basic results presented by the GAO, and so the estimates implied by that specification are also presented. Finally, the percent of joint bids, the only variable used by GAO in their basic regression that is not accounted for explicitly or by more carefully structuring the sample, is omitted because the decision to enter a joint bid is also a choice variable, and is, therefore, endogenous.[21]

The results of estimating the reduced form for the value of the high bid are shown in Table 5-1 where the numbers without the parentheses are the estimates for the coefficients associated with the variables listed in the first column. The estimates for the wildcat and proven tracts appear in columns two and three. Both reduced form models of the high bid implied by the GAO specification indicate an insignificantly positive effect of areawide leasing once other factors are taken into account based on the sign and significance of the estimated coefficients. In general, the statistical significance of the price, interest rate, and estimated value variables seems to account for the changes in the high bid. In particular, changes in price have a large estimated impact. In column two, the reduced form analog to GAO's basic equation without a dummy calendar year variable, a dollar decline in the refiners average acquisition price of domestic oil leads to a decline in the high bid of $1.4 million per tract in 1984 dollars. The total effect of areawide leasing using the GAO specification, based on the insignificance of the

Table 5-1
Reduced Form Estimates of the High Bid[a]

| Variable | Group 1 Wildcat and proven | | Group 2 Drainage and development |
	Basic	Basic with dummy	Basic
RMROV	0.89*	0.84*	0.71*
	(5.98)	(5.50)	(5.85)
AREAWIDE	603,258	2,271,621	−5,240,726
	(0.16)	(0.51)	(−0.50)
AAA	−709,591	−2,904,638*	−566,668
	(−1.62)	(−2.37)	(−0.41)
PRICE	1,360,132*	3,952,944*	973,914
	(2.16)	(1.82)	(0.60)
AVGPC	1,305,008	552,144	2,364,130
	(0.80)	(0.19)	(0.52)
CAL YR 81	—	4,152,911	—
	—	(0.37)	—
CAL YR 82	—	15,440,546	—
	—	(1.40)	—
CAL YR 83	—	22,906,695*	—
	—	(1.80)	—
CAL YR 84	—	26,476,547*	—
	—	(1.87)	—
R^2	0.36	0.38	0.37

[a] "t" statistics in parentheses; * indicates significance at the 90 % level; intercept not reported; variable definitions in the Appendix.

areawide variables, is zero. However, only approximately one-third of the variation in the high bid is explained by the model. This level of explanatory power is consistent with that obtained by the GAO (though it clearly leaves substantial variation to be explained by expanded models or improved data).

Drainage and development tracts comprise a much smaller proportion of the tracts sold; all tracts of this type receive a present value evaluation. All the drainage and development tracts in 1983 and 1984 were leased in areawide sales. Therefore, only the GAO model without calendar year dummies can be estimated for this sample because of the perfect collinearity of areawide and the calendar year dummies for 1983 and 1984. For these tracts, only the measure of value is a statistically significant predictor of the high bid. The effect of areawide sales is again statistically insignificant, though its point estimate is a decline of −$5.2 million per tract from a mean high bid in the sample of $16.5 million.

The Equality of Tract Selection and Areawide Coefficients

A more general test can be applied to the coefficients of the reduced form equation implied by the GAO model. The GAO specification models the areawide effect as altering the coefficient for the intercept but assumes that all the other coefficients remain the same. A test to determine whether *any* of the coefficients changed during the shift to areawide leasing is based on estimating the GAO specification, less the areawide dummy, on each time period separately. This test also indicates no significant effect due to areawide leasing.

The analysis of this section exemplifies how technical differences can matter. An estimate of a $7-billion loss, or an estimate of no statistically significant loss, is dependent on the method of analysis which experts can evaluate based on the technical merits of the analysis. Though the estimates reported here are improved estimates over those of GAO, not all of the issues of predicting the high bid or the number of bids have been resolved.[22] Predictions of these important measures of program performance are constantly evolving, in part spurred on by the view of groups outside the government. It is the evolution of these battles in the coming years that will largely shape the management of the OCS.

LEASE DELAY RIGHTS

Previous sections present examples of policy analysis of existing or proposed policies. Policy analysis can also be applied to existing policies to suggest changes. This section illustrates a policy analysis proposal to change the bid adequacy rules. The proposed change would allow state and interest groups seeking to defer leasing to supplement, on a tract-specific basis, the reservation price determined by the bid adequacy rules. If the supplement to the reservation price that is paid by the state and interest groups is sufficiently large, the lease is not granted in that particular sale but is reoffered, subject to the same procedure, at later sales. In effect, the state and interest groups become additional competitors in the sale by bidding for a "lease delay right" in competition with groups bidding for exploration and development rights.[23] Indirectly, the analysis presents a new argument about fair market value. That argument is whether fair market value can be received if groups who are willing to pay for a particular outcome, delay of the lease, are systematically excluded from the market.

In ways that remain to be demonstrated, this proposal of lease delay rights is directly linked to specific goals of OCS management. Lease delay rights are a market-oriented method to determine the balance between development and conservation of OCS oil and gas resources and, therefore, have the potential of being economically efficient. The process would also allow direct participation by states and interest groups, hereafter named sigroups, which could have a decisive effect on the outcome of the lease auctions. From an organizational perspective, lease delay rights may increase program revenues and decrease litigation.

Finally, the issue of income distribution can be directly addressed depending on how prior property rights are assigned in the auction. If the government would continue to act as an independent owner, states and interest groups may have to pay cash to augment the reservation price. If the government would vest varying degrees of veto power in the state or interest groups, it could create auction dollars to distribute to the involved parties.[24]

Timing and the Interests of Sigroups

A first glance at federal regulations indicates that states and interest groups could participate in the current auctions. A more careful investigation reveals that it is extremely unlikely that they would wish to participate in the current auction process.

Federal regulations state that leases may be held by: ". . . private, public or municipal corporations organized under the laws of the United States or of any State. . . . or associations of such citizens, nationals, resident aliens, or private, public, or municipal corporations, States or political subdivisions of States."[25]

This section would seem to include all of the groups that are critical of the program, and sigroups can, in fact, become qualified bidders. However, lack of bidding by interest groups in the current process may be due to several factors. One factor is that a winning bidder must be both qualified and responsible. The term "responsible" has not been codified, though one interpretation is that it is a group that can satisfy the diligence requirements of a lease; in particular, the provision that exploration, including drilling, begin during the primary term of the lease. No group that is in violation of the diligence requirements can bid on additional leases. A second and probably more important factor is that the current basis of comparison for the bids at a lease sale is the once-and-for-all-time value from extraction of the resource. Sigroups may not be willing to pay an amount equal to the once-and-for-all benefits of extraction when at most a delay equal to the primary term might be obtained. In the current process, over a time period of T years (T greater than the primary term), sigroups would have to pay [T/primary term] times the net benefits of extraction to delay the possibility of extraction for T years. For instance, if a sigroup hoped to delay development for 10 years when the primary term is five years, under current regulations the sigroup would have to pay twice the value of the tract. As T gets large, this can be many times the net benefits of extraction. This far exceeds the standard efficiency criteria that a resource be allocated to that use with the highest present value. The standard criteria imply that a one-time bid by sigroups that exceeds the present value of extraction could permanently, not just for the length of time of the primary term, postpone extraction.

Since permanent delay rights *cannot* be granted through the current auction process, the remainder of this analysis maintains the current lease structure and basic auction process. The auction process is augmented to allow sigroups to defer leasing. This is done by incorporating the cost of delaying extraction and compensating the government for the expected value of the delayed revenues, an amount necessarily less than the once-and-for-all net benefits of extraction.

Federal regulations state that "The United States reserves the right to reject any and all bids received for any tract, regardless of the amount offered."[26] It is, therefore, legally feasible to alter the bid adequacy (rejection) rules of the Minerals Management Service to allow sigroups to participate in the *rejection* process, based on their willingness to pay for a delay.

Implementing Lease Delay Rights: Bid Adequacy Rules

Consider a hypothetical sale. The issue is to define an outcome of an extended auction process to distinguish three results: (1) issue a standard lease, (2) issue a lease delay right, or (3) do not issue a lease based on current bid adequacy rules. The following discussion assumes that a high bid on a given tract exceeds the current bid adequacy criteria so that the decision is between the first two results.

In general, economic efficiency is achieved if a resource is allocated to the use with the highest present value. In the context of offshore leasing, this implies issuing a lease delay right if the present value of income from an immediate sale is less than the sum of the sigroup bids plus the expected present value of delayed income from a later sale. In other words, delay the lease if

Present value of selling today < [Present value of delayed sale + delay bid]   (4)

alternatively,

[Current income − Present value of delayed income] < Sum of Sigroup bids   (5)

The intuition in support of equation 5 is simply stated: If sigroups are willing to pay an amount equal to or greater than the expected loss of revenue from delaying the standard lease, than grant a lease delay right. The left-hand side of this equation is the lease delay cost—the minimum amount that would lead to granting a lease delay right. A negative value for the left side of equation 5 indicates that there is no need to evaluate the sigroup bids since the expected present value will be increased by rejecting the development bid and delaying the lease.[27]

The major difficulty in quantifying equation 5 is computing the lease delay cost. Two methods are discussed to calculate these measures of foregone revenues. The first method uses empirical relations between the high bid received and total government revenues including the cash bonus, taxes, and royalties. The second method uses an alternative measure of the lease delay cost (foregone revenue) that is already computed by the Minerals Management Service for some tracts.

In the first method of calculation, if a specific lease is deferred, the present value of all of the revenues from the lease has to be recomputed based on the time until the next sale. As a result, the revenues will change at least due to price changes and the further discounting of the foregone revenues. By making assumptions about the price changes, the discount rate and bidding behavior, an estimate of the present value of the foregone revenues can be determined.

The second method of calculation is the delayed mean range of value (DMROV). The delayed mean range of value adjusts for changes in royalties and excise taxes from a delayed sale based on the mean range of value of the tract as estimated by the Minerals Management Service. Hence, an alternative measure of the lease delay cost could be computed as the difference between an acceptable high bid and the delayed mean range of value.

Now, consider hypothetically how sigroups might augment the reservation price of the Minerals Management Service. Each sigroup would determine if it incurs costs up to the time of the next sale if a tract is leased. Based on these estimates, each group submits a delay bid, with the same bid submission requirements for existing bidders such as one-fifth down payment as specified in the Code of Federal Regulations.[28] The delay bids would presumably be funded by taxes or donations from members of a state, locality, fishery, or interest group.

How might the winner of the auction be determined? For those tracts where the *sum* of the delay bids exceeds the cost of delay, the high bid is rejected.[29] As a result, the delay right is granted until the scheduled date of the next sale and precludes exploration and development other than that normally allowed in the presale process. For those tracts where the sum of the delay bids does *not* exceed the cost of delay, including tracts that do not receive any delay bids, then the usual exploration and development lease is granted if the high bid passes the current bid adequacy rules. Finally, if no acceptable high bid is received, there is no cost of delay. No lease delay right is granted and no lease delay payments are required. Though procedural rules as described above can be elaborated upon, such efforts are irrelevant unless sigroups are, in fact, willing and able to bid in amounts that alter the outcome.[30]

Interest and Ability of Sigroups to Bid for Delay Rights

Economists would predict a trade-off between the investments of sigroups in an auction and nonmarket (political and judicial) activities that could lead to deferrals. Intuitively, the existence of annual leasing moratoria is probably a cost-effective action for sigroups. However, the political costs of the moratoria seem to rise and fall with environmental incidents relating to the OCS. The implementation of lease delay rights would open a sequential alternative to political action. Specifically, once political interventions such as moratoria and deferrals from sales are passed, sigroups may choose to invest in lease delay rights. Knowledge that such investments will be open may reduce the investment by sigroups in the political process until the additional return per dollar is equated across investments.[31] The abortive bid by Greenpeace, an environmental group, in an Alaska and a North Atlantic sale is a preliminary indication of interest by sigroups in the bidding process even when no administrative procedures are established for such groups. Individual states may also find it in their interest to spend revenues to delay development off their coast. The site-specific nature of a lease delay right may encourage participation in the auction by the jurisdictions that are most directly affected.

Because several sigroups may receive benefits if the lease is delayed, a free rider problem might exist. The free rider problem can occur because an individual sigroup can receive benefits if another group delays the lease. This creates a strong disincentive for any group to bid to delay a lease. One solution is to allow conditional contracts to be drawn up so that each free rider pays only if all the others pay. The mutually desired outcome succeeds when all pay, but fails if some individual group attempts to be a totally free rider. There is no guarantee that the payments agreed to in a contract will actually reflect the true willingness of the group to pay even though the desired outcome occurs. Note also that all of the sigroups, except environmental groups and trade associations, have an ability to tax. This reduces the free rider problem among members as they would be required to pay the tax.

The sigroups may also determine that they wish to pay less than their actual willingness to pay. Strategic bidding may lead to sigroups developing at a substantial cost their own expertise in evaluating the production potential of an area. The sigroups would have to decide whether their costs would be less by spending money on existing expertise or by bidding a higher, but less informed, amount for the delay bid.

Cost of Lease Delay Rights and Budgets of Sigroups

The sale of lease delay rights may be irrelevant unless the sigroups have a sufficiently large budget to expect to delay some leases. In order to analyze this issue, an average cost of delay for evaluated tracts in two specific sales was determined and then compared with the budget of possible sigroups. The data used are from sales 70 (April 1983) and 80 (October 1984) in the Bering Sea and in Southern California. Future sales were scheduled 2 and 2.5 years after the initial sale. These sales were selected as recent sales in areas of high interest to sigroups. The result was an average annual delay cost of approximately $1 million in the Bering Sea and $400,000 in Southern California.[32] Many of the tracts in the sale had annual lease delay costs less than this figure as well as those that would have a larger lease delay cost.

Table 5-2 measures the share of the annual budget of selected sigroups that would be represented by one lease delay right obtained at these average lease delay costs, with, as one saying goes, "no money left on the table." This means that the lease delay right was purchased at the lowest cost acceptable. The data indicate that a delay right would cost in the range of .001 percent to 5 percent of the budget of a single sigroup. This share of the budget of an individual sigroup potentially used on one lease delay right should be interpreted with caution. Because the cost of a delay right can be spread among other sigroups, the cost of one lease delay right may in fact be a smaller share of any particular sigroup's budget. However, this is balanced by the fact that as sigroups build up an inventory of tracts, payments would be due on many tracts.

Clearly, existing budgets are not completely available to invest in lease delay rights. However, the membership and budgets reflect the activities of the group, particularly for environmental groups. One might imagine that the opportunity for these groups to purchase tangible rights—lease delay rights—would result in an expansion of membership and budget. For instance, after an auction in which lease delay rights are granted, the acres in the lease might be allocated to the sigroups in proportion to their participation. The groups can in turn send certificates to members that they have delayed development on some number of acres corresponding to their contribution. In effect, existing groups would act as brokers for interested parties. In either case, having a tangible product to purchase may increase membership and donations.

Lease delay rights could have two significant impacts on the goals of the leasing program. The first impact is on economic efficiency as more individuals who perceive themselves as incurring costs may participate in the market. The sigroups would then

Table 5-2
Lease Delay Cost Share of Selected Sigroup Budgets

Sigroup	Population/ membership	Annual budget (million $)	Share (%) Sale 70[a]	Sale 80[b]
State of Calif.	25,174,000	$42,247	.002%	.0009%
State of Mass.	5,767,000	8,752	.01	.005
County of Santa Cruz	188,000	151	.7	.3
City of Santa Barbara	74,000	41	2.4	1.0
Sierra Club	350,000	20	5.0	2.0
Nature Conservancy	229,000	27	3.7	1.5

[a] Assumed annual delay cost = $1,000,000 for sale 70.
[b] Assumed annual delay cost = $ 400,000 for sale 80.
Sources: States: U.S. Dept. of Commerce, *Statistical Abstract of the U.S.,* 1985, pp. 272–278; Cities and Counties: U.S. Dept. of Commerce, Bureau of the Census, *County and City Data Book;* Organizations: Membership: Gruber, K., ed., *Encyclopedia of Associations, 1986,* 20th ed. (Detroit: Gale, 1986), Budgets: annual reports, 1984.

be allowed to allocate their budgets between market and political processes as methods of achieving their ends. The auction outcome is then determined impartially based on those that have the highest willingness and ability to pay whether the outcome is exploration and development or delay. Expanded access to the auction process may also result in improved perceptions of fairness of the program (though that is not an explicit goal of OCS leasing).

The second impact occurs when the government is compensated for expected, foregone revenues if lease delay rights are, in fact, granted. The income from the lease delay rights replaces revenues foregone above some determined reservation price. Lease delay rights would be neutral with respect to expected government revenues based on the reservation price. However, program revenues may actually increase relative to an auction when lease delay rights are not possible. This is because the bidding strategy of development firms may be altered by the possibility of a higher, aggregate reservation bid. In effect, a new set of bidders have entered the auction to increase the competition on selected tracts.

Fair market value has been said to be the politician's dream: one can criticize offshore leasing while also being in support of increasing government revenues. In practice, fair market value remains an elusive quantity, yet the policy analyses of the rate of return, of the effect of areawide leasing, and the role of lease delay rights illustrate results that inform management in their search for fair market value.

Chapter 6

Aggregate Analysis: the Pace of Leasing and Financial Accounting

Aggregate analysis of management by the Minerals Management Service involves analyzing the timing, size, and location of lease auctions—the pace of leasing—and the aggregate financial evaluation of the leasing program. These analyses are in contrast to analyses in earlier chapters that use tracts, prospects, or individual spills as the unit of analysis.

The first section has three parts related to the pace of leasing. The first part provides institutional background on the pace of leasing. The second part is a survey of the theories considered applicable to the pace of leasing, and the third part presents the concepts of privatization and federalization as competing organizational principles. The second section of this chapter uses financial accounting to provide an aggregate analysis of the pace of leasing as well as information on the several uses of funds collected from leasing. The final section analyses the statistical relationship between the 1987–1992 five-year plan and the data generated by the Minerals Management Service.

THE PACE OF LEASING

The pace of leasing is a policy decision regarding the number of lease sales and the area available for lease. The pace of leasing is the key determinant of the supply of leases that affects in different ways the achievement of virtually every goal for OCS management: expeditious exploration, the balancing of environmental effects, alternative uses, socioeconomic effects, and the receipt of fair market value. However, as economists are fond of saying, it takes both blades of a pair of scissors to cut. The pace of leasing only affects the supply of leases during a period of time, the actual effects are the result of the auction process that merges the supply of leases with the demand for leases by industry. The public documents that define the pace of leasing— the five-year plan and the various sales announcements—are also opportunities for state and interest group consultation as discussed in Chapter 2.

Because no agreement on a method to determine the appropriate pace exists, the pace of leasing is a contentious policy issue. This is not just a problem for the public at large but also for the academic community where the difficulty in modeling uncertainty in its many dimensions, the exploration and production choices of oil and gas firms, and the role of OCS resources in a worldwide system over time all contribute to the lack of unanimity in approach.

Over the years, the pace of leasing as measured by the number of sales, acres offered, and acres leased has changed substantially. Table 6-1 shows the number of sales, the acres offered, and the acres leased from 1972 to 1986.

It is clear from Table 6-1 that the number of acres offered and leased increased substantially from 1983 to the present. More specifically, though there have been some changes in the definition of the sale area (for instance, in the early 1970s sales were organized by state in the Gulf of Mexico), it is clear that a steady number of sales has occurred and is projected to occur in the Gulf of Mexico. Other areas, ranging from the Pacific to the Atlantic to Alaskan waters, have been leased more sporadically.

Since the 1978 amendments to the OCS Lands Act, the basic pace of leasing is determined by the five-year plan mandated in those amendments. The focus of the five-year plan is on 26 planning areas, 15 of which are in Alaska. The pace of leasing is further refined for each particular sale through the public review and comment process

Table 6-1
Pace of Leasing—1972 through 1992

Year	Gulf of Mexico			Other Areas		
	Sales	Acres offered	Acres leased	Sales	Acres offered	Acres[a] leased
		Millions			Millions	
1972	2	1.0	.8	0	0	0
1973	2	1.7	.5	1	.8	.5
1974	5	5.0	1.8	0	0	0
1975	3	6.0	1.3	1	1.3	.3
1976	2	1.0	.4	2	1.9	.9
1977	1	1.1	.6	1	.8	.5
1978	3	1.8	1.0	1	1.3	.2
1979	2	1.2	.8	4	2.3	.9
1980	2	1.4	1.0	1	1.2	.2
1981	2	2.2	1.3	5	5.5	.9
1982	2	1.9	.9	3	5.6	1.0
1983	3	71.2	5.4	5	48.7	1.2
1984	3	115.3	5.1	3	38.9	2.2
1985	3	87.0	3.6	0	0	0
1986	2	58.7	.7	0	0	0
1987	2	59.7	3.4	0	0	0
1988	3	107.9	4.8	3	49.5	3.2
Proposed Sales						
1989	2			7		
1990	2			6		
1991	3			4		
1992	1			7		

[a] Sales include frontier and supplemental sales.
Sources: Minerals Management Service, *Final Five Year Plan; Federal Offshore Statistics; Outer Continental Shelf Lease Offering Statistics,* May 15, 1989, prepared by Eileen Swiler, MMS-Gulf of Mexico.

described in Chapter 2 and through the inevitable unfolding of events. For instance, a sale planned for the Gulf of Alaska in 1990 was delayed by a decision to wait for new environmental information that may result from the *Exxon Valdez* oil spill.

The planned pace of leasing as defined in the five-year plan for 1987–1992 is broadly based on making available for lease those areas that are estimated to have positive net social values. A positive net social value for an entire area results from social costs, as discussed in Chapter 4, that are less than the social benefits of production. Social benefits include the private expected value plus government payments in the form of taxes and royalties. Those areas with the largest positive net social values have sales scheduled most frequently (as often as once a year in the Gulf of Mexico). Other areas with lower but still positive net social values may have planned lease sales as infrequently as once in five years. Many of the remaining areas that do not have positive net social values at any of the range of oil prices investigated by the Department of the Interior are scheduled for one sale during the five-year plan, whereas some do not have any planned lease sales.[1] The Department of the Interior typically justifies holding a sale in some areas that do not have positive net social value by considering the importance of the area for the potential discovery of oil and gas as a mandated goal.

The current procedure for determining the schedule of lease sales has evolved through statutory interpretations by the Minerals Management Service and litigation. Litigation from the first two five-year plans led a federal district court to rule that a five-year plan will be accepted as long as three standards are not violated. These three standards are that the Department of the Interior, through the decisions of the secretary, must (1) make factual decisions on the basis of substantial evidence, (2) make policy decisions that are neither arbitrary nor irrational, and (3) must effectuate the intent of Congress in the administrative interpretations of OCS Lands Act Amendments.

In many areas this leaves substantial discretion to the secretary to conduct an analysis and apply a decision criteria within the general guidelines given by Congress. In particular, the court determined that "within the section's 'proper balance' there is some notion of 'costs' and 'benefits,' recognizing that 'costs' in this context must be a term of uncertain content to the extent it is meant to stand for environmental and social costs."[2] The further use of cost benefit analysis as an important but not unique guide to the frequency of leasing was upheld without placing constraints on the area to be offered. The court allowed the Department of the Interior substantial leeway in determining what methods to use. The court further stated: "Where existing methodology or research in a new area of regulation is deficient, the agency necessarily enjoys broad discretion to attempt to formulate a solution to the best of its ability. . . ." The following section surveys some variations of the economic theory that are (or could be) used to justify the pace of leasing.

The first component of a supporting economic theory focuses on the highly aggregated measures of net social value for each of the 26 planning areas. The current rationale for the frequency and location of lease sales is most closely analogous to a theory of resource supply from 26 large deposits of different quality where increases in production lead to increases in cost in each area. The extraction policy that results in economic efficiency is when extraction occurs first from the deposit with the lowest average cost. Other deposits are exploited concurrently if the added cost of production from the first deposit—the marginal cost—exceeds the cost of starting production on another deposit. In effect, as the frontier of oil and gas development in the Gulf of

Mexico has moved into deeper water with resulting increases in cost, it becomes worthwhile to exploit deposits in other areas that were previously too costly to produce (recall that costs are defined broadly as in the court's opinion). The Minerals Management Service could be said to be making available (at an aggregate level and at discrete intervals) the areas that allow oil and gas companies to achieve economic efficiency.[3]

There is a second component to the economic criteria. Not only should leasing proceed from low cost to high cost, but the timing should ideally be such that, measured in today's (discounted) dollars, the net benefits are not higher if the sale is shifted to any other time period. To date, this component is only partially addressed on a tract-specific basis through the use of the delayed mean range of value in the bid adequacy criteria.

Economists are still uncertain whether energy development companies act in a way that is consistent with economic efficiency,[4] but it is clear that leasing can take place too rapidly or too slowly. The analysis of information externalities in Chapter 3 is a case where exploration can proceed too slowly. Other cases can exist where individual companies may rush to bid for a tract once it is expected to have a positive private value (not necessarily a positive social value) and perhaps before the social value is as large as possible. Whereas it is possible to use the current Minerals Management regulations for diligent development, royalty rates, and the minimum bid to achieve economic efficiency at the actual time of the sale, it is doubtful that current policies achieve this objective.[5] The reason is that fine tuning of the regulations would be very data intensive, subject to considerable uncertainty, and cumbersome.

There are additional theoretical complications that can lead to modified development rules in order to achieve economic efficiency. Some of these complications involve the uncertainty of the size of the resource stock,[6] the effect of uncertain prices,[7] the effect of depletion,[8] the value of partially but irreversibly developing a resource if that development provides information,[9] the effect of a backstop energy source at either a known or unknown price,[10] and the strategies that countries might pursue if they are significant players in the world energy market.[11] Each of these theories leads to different definitions of economic efficiency. Economists are still seeking either the special model that, for instance, leads to accurate predictions of OCS extraction, or a sufficiently general model of resource extraction that the OCS problem becomes a well-defined special case.

These additional complications of theory have to be balanced with the added analytical complexity and difficulty of implementation. In that regard, Congress originally gave the Secretary of the Interior nine months to construct a five-year plan (though the 1987–1992 plan required more than two and one-half years). More recently, organizational efforts within the Minerals Management Service for the 1992 plan began in 1989.

In fact, however, much of economics is devoted to showing that markets left to themselves can achieve an economically efficient outcome. In particular, some representatives of industry and some academics argue that complete divestiture of OCS mineral rights (even at no cost that would, however, violate the congressional mandate for fair market value) could achieve economic efficiency. Such divesture would be an example of a privatization policy as discussed in Chapter 2.[12] A market, however, achieves economic efficiency only if the buyers and sellers consider all of the relevant costs (including the environmental costs discussed in Chapter 4) and if the rights to the resources can be bought and sold in a market. Whereas some congressional goals may not be considered by firms, several are, at least in part. Firms must incorporate some concern for the environment and alternative uses of the ocean such as fisheries because OCS operators are liable for damage to the environment and to fisheries, a topic discussed in Chapter 4.[13]

In addition to privatization, another management approach to the pace of leasing could be to federalize management by divesting the OCS to the states. However, the economic concept that a buyer and seller must consider all the costs and benefits in order for a market to be efficient has a parallel in organizational incentives. In broad terms, the current allocation of benefits and costs to different organizations is not consistent with these organizations agreeing on an economically efficient pace of leasing. Table 6-2 indicates the differing incentives for different areas of the OCS; the state offshore lands totally controlled by the states, the 8(g) lands that extend for an additional three miles in which government revenue is shared between the state and the federal government, and finally the OCS lands themselves.

There are clearly different perceived strategies and different incentives for each type of organization. Texas and California manage their state offshore lands very differently, a fact also reflected in that Texas has chosen not to participate in the Coastal Zone Management Program, whereas California has a very active coastal zone program. These differences may reflect different costs and benefits associated with development in each area.

On an informal basis, there is also evidence for the interest of states in expanded revenue sharing from the OCS lands. Finally, there are small costs incurred by states when they resist oil and gas development on the OCS lands, particularly beyond the 8(g) lands.

In contrast, whereas managers in the Department of the Interior must consider the regional distribution of benefits and costs, their most direct incentive from leasing is the receipt of revenue into the federal treasury. Economists would find it very surprising if these different organizations agreed on the same pace of leasing. The following section is an analysis of OCS revenues received and expenses paid by the federal government, and the distribution of revenue to the states through several mechanisms.

ACCOUNTING FOR MANAGEMENT

Accounting serves as the aggregate, annual yardstick by which the pace of leasing and the receipt of fair market value can be measured. Accounting is particularly important

Table 6-2
Relative Organizational Incentives

Organization	Strategy	State Lands	8 (g) Lands	OCS Lands
Texas	lease early	all income	27% income	no income
		all cost	all cost	all cost
California	don't lease	all income	27% income	no income
		all cost	all cost	all cost
Dept. of	increase	no income	73% income	all income
Interior	leasing	no cost	no cost	no cost

to the management of the Outer Continental Shelf lands because revenues from offshore oil and gas production must first be apportioned between the federal government and private firms, and then federal funds must be apportioned between several major accounts of the government and the coastal states. To further complicate accounting for OCS leasing, the Minerals Management Service is also the collection agency of the federal government for *all* mineral revenues whether offshore or onshore.

The Minerals Management Service presents its revenues and expenses annually to the Office of Management and Budget in the executive office and to the Congress for budgetary overview. This section expands on these data to create integrated financial statements as reflected in an income statement, in a funds statement, and in a balance sheet.

The presentation of an integrated income statement and balance sheet is an everyday occurrence in the private sector. However, federal government accounting is fundamentally different from private sector accounting and no such integrated statements are prepared. The reason is that governmental accounting systems are based on fund accounting, which devotes little attention to long-lived assets but is instead oriented toward yearly appropriation and revenue cycles. Fund accounting is suitable for many parts of the federal government that are based on the yearly provision of services that do not require substantial capital assets. The fund system as used in the federal government is less appropriate when long-lived assets are the core of the management problem such as that faced by the Minerals Management Service.

The Minerals Management Service is, in many respects, like a real estate holding company of a large conglomerate. Like a real estate holding company, it makes contracts based on land holdings, in particular oil and gas leases. It then collects the money and turns it over to the treasury of the conglomerate. In other respects, the Minerals Management Service is unlike a real estate holding company. It does not acquire land and its objective is not to maximize the wealth of the conglomerate. Nonetheless, the income from the lease contracts would place the Minerals Management Service approximately 80th among the 500 largest manufacturing companies in the United States based on sales.

An income statement and a balance sheet, readily understood by analysts, provide important insights into the management of the Minerals Management Service and the larger federal government of which it is a part. In particular, the analysis highlights the transfer of long-term assets into current income. However, not all questions are answered by such an analysis. For instance, the analysis of financial statements in the private sector is often relative to another company in the same industry. Comparisons—such as rates of return—are difficult to make for a government agency.

A complex bundle of rights are associated with ocean resources. The federal offshore energy program can be thought of as analogous to managing the mineral rights below a variable depth onshore (since the water depth varies offshore). The analogy is such that the Minerals Management Service can assign the variable depth subsurface mineral rights, but other agencies, including the states, affect or control the rights to resources in the water column and surface. For instance, the Minerals Management Service and its lessees are liable for damage to these water column rights in the sense that compensation funds for fishermen and for individuals damaged by oil spills are funded by the mineral resource revenues. From an economy-wide perspective the assets of the surface and the water column including fisheries, transportation, recreation, and waste assimilation are very valuable. However, this section focuses on an accounting of the subsurface energy assets and the potential liabilities to assets managed by other agencies or owned by private individuals.[14]

The Income Statement: Revenue

The Minerals Management Service is in the financially enviable position of earning revenue from land obtained "free" but then leased at auction. The primary revenues— the bonus payments on new leases, royalties from existing leases with production, and rent on leases that do not have production—have been discussed. These revenues represent a shift of assets held either as existing leases or as unleased lands into current income. The present value of long-term assets is reduced by the amount of revenues.

The fiscal year 1987 revenue statement presented in Table 6-3 also reflects large distributions from escrow accounts for past bonus, royalty, rental, and interest income from tracts in the disputed 8(g) zone. In addition, two other revenue items are listed in the income statement in Table 6-3, though they are not officially listed as receipts to the Minerals Management Service. These two revenue sources are associated with a tax on production and a lease tax that are allocated to the Offshore Oil Pollution Fund (managed by the Department of Transportation) and the Fishermen's Contingency Fund (managed by the Department of Commerce). These are essentially insurance funds— paid out of revenues from the OCS—to compensate groups potentially damaged by oil and gas operations.

In 1987 bonus and rental payments for lands newly moved from an unleased category into a leased category totaled $419 million. The release of escrowed bonus and rental payments was $1.2 billion. Royalties from producing leases totaled $2.4 billion, of which $102 million were funds released from escrow. The several insurance funds collected $13 million with an additional $7 million earned in interest earmarked for those funds (included in the fund totals). Other interest earned from the funds held in escrow was $903 million. In total, the offshore revenues of the Minerals Management Service were $4.9 billion in 1987.

The Minerals Management Service and the managers of the various funds also incur expenses for evaluating resource holdings, conducting auctions (including environmen-

Table 6-3
Revenue—FY 1987 (millions of dollars)

	Current	Escrow release	Total[a]
Bonus and rental	419	1,185	1,603
Royalties	2,316	102	2,418
Interest	0	903	903
Pollution Fund	18	0	18
Fisherman's Fund	2	0	2
Total	2,755	2,190	4,944

[a] Details may not add to total due to rounding.
Sources: Minerals Management Service, *Federal Offshore Oil and Gas Statistics*; U.S. Office of Management and Budget, *Appendix: Budget of the United States, 1989*; Minerals Management Service, *Offshore Oil and Gas Leasing/Production, Annual report, FY 1987.*

tal impact statements), monitoring offshore operations for safety, collecting and monitoring royalty payments, and administering claims. The Minerals Management Service is also responsible for monitoring the collection of royalties from onshore mineral leases on federal and native American lands. Since onshore lease collections for the entire 50 states comprise about 20 percent of the total collections, the reported costs of royalty management and general administrative expenses are reduced proportionally prior to the construction of Table 6-4.

The major expense categories in Table 6-4 are leasing management, resource evaluation, royalty management, regulation, and general administration. Royalty management and regulation are associated with the postleasing activities of the Minerals Management Service. Regulation refers primarily to the inspection and regulation of operations. Leasing management and resource evaluation refer primarily to the preleasing activities associated with research for environmental impact assessments and evaluations for determining whether a high bid exceeds a threshold of fair market value. Claims on the liability funds and their administration were relatively small ($1 million) with all of the claims paid through the Fishermen's Contingency Fund. No claims were paid through the Oil Pollution Fund in 1987, and, in fact, no claims have been paid to date since the fund's inception in 1978 (claims have been paid directly by the responsible parties).

In private sector parlance, the Minerals Management Service would be seen as a cash cow. The government spent $152 million to net $4.8 billion in revenue. In the actual government accounts, no measure of net income—revenues minus expenses—is computed because the majority of revenues are defined to be "undistributed proprietary funds," which are not used to offset costs as possibly implied by a net income calculation. Calculating net income and tracking its disbursement does, however, provide insight into the uses of the money. Also, since the net income is generated by liquidating a long-term asset, further distributions of net income depend on the size of long-term assets on the balance sheet.

Table 6-4
Expenses FY 1987

Categories	Million dollars
Preleasing	
Leasing Management	
Environmental studies	23
Leasing and environmental assessment	16
Resource Evaluation	26
Postleasing	
Royalty Management	37
Regulation	28
Other	
General Administration	21
Claims and administration of funds	1
Total	152

Sources: Appendix: Budget of the United States Government: FY 1989, op. cit.; Minerals Management Service Information office, detailed tables, Minerals Management Service Budget: FY 1989.

Funds Statement

In the private sector, analyzing the distribution of the net income is equivalent to a funds statement in the private sector—a title that is even more appropriate when applied to government accounts. The Minerals Management Service is funded entirely out of General Fund allocations to the Department of the Interior. Although this source of funds for the Minerals Management Service is somewhat uninteresting, the uses of funds are of substantial interest and are presented in Table 6-5.

In 1987 of the approximate $4.8 billion of net income, 80 percent went into the General Fund out of which most government activities are funded; 17 percent went to the Land and Water Conservation Fund managed by the National Parks Service for the purpose of buying new land and providing recreational facilities. The Land and Water Conservation Fund distributes funds to several agencies and the states; however, it is not authorized to spend all of the money transferred to it. This practice led to a $5-billion unappropriated balance in the fund at the end of 1987, though the money is actually spent for other, authorized purposes of the federal government, a topic discussed in more detail later in this chapter. The National Park Service also received $150 million of the net income for the Historic Preservation Fund, which also had an unappropriated balance of $1.1 billion at the end of 1987. Both the Oil Pollution Fund and the Fishermen's Fund collected more in revenue than was used in administration or paid out in claims. In the case of the Oil Pollution Fund, the excess of income was used to buy government securities, which show up as a change in assets. Some money was transferred to the Coast Guard (the administrators of the fund). In the case of the Fishermen's Contingency Fund, the excess of income was carried forward to the next year's budget.

The government appears to have chosen to directly divert less than 25 percent of the income from long-term assets to purchase other long-term assets (the Land and Conservation Fund and the Preservation Fund), and of these diverted funds, only one-quarter was actually expended. All other money generated by the liquidation of a long-term asset is used as a one-time transfer to support current expenses or unidentified capital improvements from the General Fund. This is not necessarily a bad strategy as the interest on further federal debt is avoided if government spending cannot be reduced. However, these transfers are a one-time phenomena whose use in funding permanent programs should be questioned. Counterbalancing the general purpose transfers from

Table 6-5
Funds Statement: FY 1987

Distribution of net income	Million Dollars
General Fund	3,800
Historic Preservation	150
Land and Water Conservation	824
Oil and Fishermen's Fund increases	18
Total	4,792

Sources: Minerals Management Service, *Federal Offshore Oil and Gas Statistics*; U.S. Office of Management and Budget, *Appendix: Budget of the United States, 1989.*

the Outer Continental Shelf to the General Fund are the earmarked transfers to the Land
and Water Conservation Fund and to the Historic Preservation Fund.

The two special funds—Land and Water, and Historic Preservation—are the most
visible way in which the benefits of offshore development have been shared among the
50 states and the territories. Because the Land and Water Conservation Fund is larger
than the Historic Preservation Fund, this section focuses on the former even though the
broad outline of funding and distribution of benefits applies to the latter fund as well.

Beginning with legislation in 1965, the Land and Water Conservation Fund primarily
receives revenues from offshore oil and gas leasing. The amount actually spent from
the fund has varied widely from just a few million dollars in 1965 to a peak of $800
million in 1978. Legislation required that at least 40 percent of the expenditures from
the fund be used for federal land acquisition. Thus the fund has helped to expand the
recreational opportunities provided by the federal government through the many divi-
sions of the National Park Service, the Bureau of Land Management, and other government agencies. Another portion of the fund supports recreational opportunities at
the state and local government levels. The history of this funding is shown in Figure
6-1.

Though the amount going to the states has declined over the years, the fund has

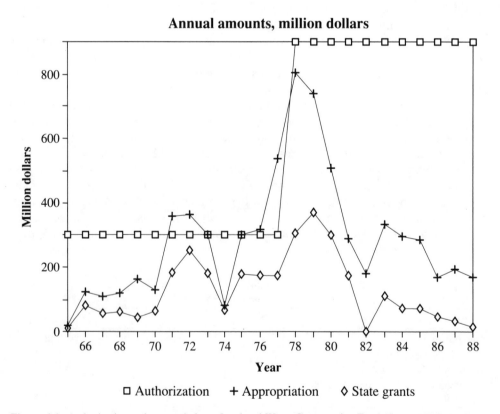

Annual amounts, million dollars

□ Authorization   + Appropriation   ◊ State grants

Figure 6-1. Authorization and appropriations, Land and Water Conservation Fund (*Source:* U.S. National
Park Service, *Annual Report to Congress: Land and Water Conservation Fund Grants-in-aid Program,*
Fiscal year 1987.)

provided baseball diamonds, camping sites, swimming pools, picnic areas, and zoos—virtually anything that a city, county, or state parks and recreation department might undertake. In order to make use of the fund, the local governments must also invest at least 50 percent of the cost of the project.

Geographically, the money has gone to local areas all across the United States. Eighty-eight projects have been funded in New York City alone, and Massachusetts has received over $78 million (an amount exceeded by California, Florida, Illinois, Michigan, New Jersey, New York, Ohio, Pennsylvania, and Texas). Figure 6-2 indicates the widespread use of the Land and Water Fund at the state level.

The money reinvested in parks, recreation, and historical preservation is the identifiable legacy of government revenues from the Outer Continental Shelf. As of 1989 legislation has been proposed to stabilize appropriations for these purposes at a higher level. The legislation follows a recommendation from the President's Commission on Americans Outdoors and is championed by Congressman Morris Udall (D) of Arizona.[15] Under the proposed legislation, essentially the same amount of money from

Figure 6-2. Cumulative appropriations to states, Land and Water Conservation Fund (*Source:* U.S. National Park Service, *Annual Report to Congress: Land and Water Conservation Fund Grants-in-aid Program,* Fiscal year 1987.)

offshore leasing that is currently placed in the Land and Water Fund and the Historic Preservation Fund would go into a new fund called the American Heritage Trust Fund. The money from this new fund would be available for appropriation without prior authorization by Congress. Proponents of the American Heritage Trust Fund feel that by bypassing the authorization process, appropriations for this use of OCS revenue would be more stable. Opponents of the proposal argue that because of the financial condition of the government, revenues from OCS leasing should go to support general expenses instead of being committed to recreational development.

The issue of what to do with money from developing the Outer Continental Shelf is likely to remain political. The money is the reward from managing the resources and can either be squandered or reinvested wisely just as can other sources of government revenue. In any event, revenue management is one of the least prescribed aspects of OCS management.

The Balance Sheet

The revenues for the funds discussed above are generated by a shift of assets from long-term assets to current income. The balance sheet categorizes the estimated present value available to the government for further transactions of this type. Estimates of the assets available are again based on the principles discussed in Chapter 3. In particular, the estimates are based on the technological and economic conditions at a particular time. The balance sheet in Table 6-6 focuses on the asset value of the mineral holdings and the long-term assets of the Oil Pollution Fund. Any unobligated balance of current budget authorization and the value of assets such as equipment are not reported.

The first long-term asset is government securities held by the Oil Pollution Fund. These totaled $103 million as of the end of 1987. These securities are held for potential claims. Since the targeted statutory balance of this fund is between $100 and $200 million and the balance falls within that range, new payments into the fund could be suspended with congressional action. Alternatively, oil spills from the *Exxon Valdez* and elsewhere in the late 1980s may lead Congress to increase the amount in the fund.

The vast majority of the assets of the federal offshore energy program are the cur-

Table 6-6
The Balance Sheet

Long-Term Assets	Million 1987 Dollars
Government Securities	103
Leased Tracts	
Developed reserves	18,560
Undeveloped reserves	2,318
Unexplored	6,562
Unleased Tracts	
Unexplored	9,800
Total	37,343

Source: Adapted from D. Rosenthal, M. Rose, and L. Slaski, "The Economic Value of the Oil and Gas Resources on the Outer Continental Shelf," *Marine Resource Economics,* 1989.

rently and potentially leasable lands of the Outer Continental Shelf. In a recent article, Rosenthal, Rose, and Slaski estimated the present value of government receipts that would result based on the current technology, a forecast of prices, and leasing rates until the year 2016.[16] Clearly, a large degree of uncertainty—even more than that faced by a private petroleum company—can be found in these estimates. Whereas both the Minerals Management Service and a private petroleum company face uncertain prices and uncertain geology, the government's problem is compounded by the fact that the auction procedure makes private companies the agents in achieving the objectives of the government.

Various categories of uncertainty are reflected in the balance sheet. Conceptually, Rosenthal and colleagues divide the assets managed by the Minerals Management Service into four categories: leased tracts with developed reserves, leased tracts with undeveloped reserves, leased tracts that have not yet been explored, and unleased tracts. The bonus revenue has already been received on the leased tracts but is a future income on the unleased tracts. Each of the categories of leased tracts has different costs that affect the calculations. In general, however, each category of asset, whose value on the balance sheet is the present value of the lease to the government, incorporates the expected time stream of royalty, rental, and, where appropriate, bonus payments. The values are discounted at 8 percent and incorporate an assumed 1 percent price increase per year from a base of $18 per barrel in 1987.

In present value terms, the expected value of already developed reserves comprise almost half the asset value and are estimated to be worth almost $19 billion. Approximately 25 percent of the asset value is the expected present value of unleased tracts worth almost $10 billion. This contrasts with almost $90 billion (nominal) already received by the Department of the Interior from OCS leasing.

The balance sheet is, however, sensitive to several sets of assumptions, and the actual dollar amounts realized are likely to be very different due to the large variance associated with these expected values. One set of assumptions underlying the balance sheet refers to price and discount rates, which are discussed by Rosenthal et al.[17] They find that the total value of leases would increase to $51 billion if a 6 percent discount rate is used or would increase to 61 billion if the 1987 oil price had been $26 dollars per barrel.

The importance of these assumptions is underscored by comparing the relative value of leased and unleased tracts to the estimates of Boskin et al. (whose data appear in Table 1-1),[18] who assumed that prices would rise at the rate of discount; their estimates of the present value of unleased resources as of 1981 indicated that the value of unleased resources was over seven times the present value of leased resources. The intervening period to 1987 has seen large-scale shifting from the unleased to the leased category, and prices and probabilities of discovery have changed substantially. However, the difficulty in reconciling estimates of balance sheet items prepared in different ways at isolated points in time indicates the potential value of a regularly prepared data series.

Financial Analysis

Analysts in the private sector often rely on several ratios from income statements and balance sheets to provide a summary of the firm's position. These summary ratios are not meant to replace detailed analysis but do provide some indications of the aggregate financial status of a company.

Many of the ratios relevant to the private sector are not relevant to the accounts as presented here. For instance, the balance sheet does not contain liabilities because the Minerals Management Service received its title to the land free. In effect, the land is a contribution to stockholders' equity from the shareholders—the citizens of the country. As a result, the notion of stockholders' equity—assets minus liabilities—is simplified to be total assets. Ratios such as the debt to equity ratio or current assets to current liabilities are not relevant. However, several ratios such as the asset earning rate and some operating ratios are relevant; see Table 6-7.

Operating ratios emphasize in proportional terms the importance of the various revenue and expense items. In fiscal year 1987, total royalty payments were almost 50 percent of the total revenue as compared to 32 percent due to bonus and rental payments. (A review of Table 6-3 indicates that these ratios are somewhat different when escrowed income is not included.) The operating cost ratios for the major items listed range from 15 to 24 percent of total expenditures. A time series of ratios of this kind can reveal major shifts in policy if shifts in expense categories are required as well. One example of such a shift might be if increased effort is devoted to postlease inspection and regulation instead of to prelease resource evaluation.

The asset ratios require careful interpretation as they are ratios of present value items. For instance, the all asset earnings ratio equal to .13 indicates that in present value terms the government received 13 percent of the total present value of its remaining offshore energy assets in 1987. In nominal terms, however, the government will receive a substantially greater amount, perhaps augmented by any difference between the expected present value of the leases and their realized value.

Since shares of stock are not sold in the Minerals Management Service, one may think that a price-earnings ratio is not computable. Some insight can be gained, however, by assuming that a perfectly competitive market for shares of stock would value the Minerals Management Service at the present value of its assets. In that case the "price" (value of shares of stock) to earnings ratio would be the ratio of total assets to net income. In 1987 this value was 8—somewhat low, but within the range of common price-earning ratios on the New York Stock Exchange.

Table 6-7
Financial Ratios

Ratio	Definition	Value
Operating Ratios		
Bonus	Bonus/Total revenue	.32
Royalty	Royalty/Total revenue	.49
Environment	Environ Stud/Total expenses	.15
Evaluation	Evaluation/Total expenses	.17
Collection	Royalty man./Total expenses	.24
Asset Ratios		
Leased asset earnings	Total revenue/Leased assets	.18
All asset earnings	Total revenue/Total assets	.13
Price-earnings	Total assets/Net income	8

The financial statements presented in this section become increasingly useful if they would be compiled annually. While the components for the income statement are compiled annually, the balance sheet as presented in Table 6-6 is not an ongoing computation in the Minerals Management Service. While one could adopt an inventory adjustment approach to the balance sheet, revaluations due to price changes, new discoveries, or lack of discoveries require the detailed information possessed only by the Minerals Management Service.

Finally, the financial statements presented in Tables 6-3 through 6-7 encourage one to investigate the tradeoffs between holding long term assets and using the revenue generated by their sale. A private oil firm, wishing to stay in the same business, would use net income to discover more oil or to conduct research and development on alternative sources. A conglomerate, more akin to the situation in government, may wish to diversify its holdings into alternative long-term assets such as through the Land and Water Conservation Fund and the Historic Preservation Fund. Based on the figures presented in these tables, the majority of the present value of the long-term assets is already committed to leasing status and the transfer of funds from long-term assets to current income is a relatively temporary phenomenon. Furthermore, income statements after 1987 would be less influenced by transfers from escrow accounts.

Previous sections of this chapter discuss the pace of leasing in terms of the number of sales and how the income and expenses associated with sales can become part of a standard financial analysis. The following section returns to the issue of the number of sales specified in the current five-year plan and the legal balancing requirements contained in the Outer Continental Shelf Lands Act.

DOES ANALYSIS MATTER?

Some critics contend that the extensive analysis conducted by the Minerals Management Service in the five-year plan is not the actual determinant of the leasing schedule. For instance, industry ranking, an acceptable balancing variable, might be viewed as receiving virtually exclusive attention. The five-year plan provides a relatively unique public document on which to apply quantitative policy analysis for the significance of the data used in determining the leasing schedule. Though neither the complex decision-making process in the Department of the Interior nor the final weighing of considerations by the secretary can be observed, one can observe the resulting plan and the analytical inputs. The issue is whether the analytical inputs are statistically related to the resulting plan and whether the analysis was policy significant.[19]

The distinction between statistical significance and policy significance is an important one. This issue was recently addressed by McCloskey who discussed tests of significance from regression analysis.[20] McCloskey stated:

> Roughly three-quarters of the contributors to the *American Economic Review* misuse the test of significance. They use it to persuade themselves that a variable is important. But the test can only affirm a likelihood of excessive skepticism in the face of errors arising from too small a sample. The test does not tell the economist whether a fitted coefficient is large or small in an economically significant sense.[21]

The following analysis considers that usual tests of statistical significance are necessary but not sufficient to address the question of whether analysis matters. The policy significance will be determined by the change in a planning variable estimated to be necessary to cause an increase of one in the expected number of sales in an area. The results of the analysis indicate that the input variables to the planning document are statistically significant but policy insignificant.

Any statistical model of a process that required two and one half years and generated more than eight inches of supporting documentation is necessarily a simplification. A simplified model of this process is that the secretary attempted to choose a leasing schedule that would maximize his satisfaction. The expected satisfaction to be gained by holding lease sales in an area is assumed to be based on the information about each area provided in the planning document. The data are not constructed on the basis of the number of sales to be held in an area so the data can be considered exogenous factors.

The problem of determining the number of sales scheduled for an area conditional on previously determined base line data differs slightly from a standard regression model such as that used for areawide leasing in Chapter 5. In the standard regression model of the previous chapter, a researcher estimates values of a coefficient that links causal variables to a continuous measure of the outcome.

The current problem differs from this standard problem because there are a discrete number of outcomes (sales) *and* the outcomes have a quantitatively meaningful numerical ordering.[22] A statistical model of this process must represent nonnegative whole numbers. The Poisson distribution has been used to represent such a process.[23] Ultimately, however, the result is similar to the standard regression model in that the analysis quantifies the conditional average of a variable. In this case, the logarithm of the average number of sales was modeled as being a linear function of the explanatory variables. The logarithmic form of the mean imposes the theoretical restriction that the average be positive.

Data and Hypotheses

The five-year plan presented a variety of data to the secretary. Some of the most prominent data were the measures of net social value for each planning area as discussed at the start of this chapter. Table 6-8 presents data on net social value as they appeared in the five-year plan; Figure 6-3 links the data to their geographic location. Other data contained in the plan included groups of planning areas that the Department of the Interior developed from the range of net social value estimates, industry rankings, environmental sensitivity, and the number of written comments received about each area in the five-year plan. It is also possible to construct a variable to measure the probability of litigation in an area based on the number of past sales that have been litigated. These data were used as conditioning factors that might shift the average.

The hypothesis tests for the statistical significance of the data in the five-year plan are relatively uncomplicated. What variables are statistically significant determinants of the number of sales? In particular, are analytical measures such as net social value significant? Or are the indicators of political effect (such as the number of comments) significant? More importantly, measures of policy significance are defined by the change

Table 6-8
Estimated Social Values

Range of estimated net social value of total production of unleased, undiscovered OCS oil
and gas leasable resources as of mid-1987

Planning Area	Column 1 Variation of Estimated Net Economic Value of Leasable Resources ($ 1987 Millions)		Column 2 Variation of Estimated Social Cost of Producing Leasable Resources ($ 1987 Millions)		Column 3 Variation of Estimated Net Social Value ($ 1987 Millions)	
	(1A) Low price case	(1B) High price case	(2A) Low price case	(2B) High price case	(1A-2A) Low price case	(1B-2B) High price case
Central Gulf of Mexico	9,432	$31,236	42	42	9,390	31,194
Western Gulf of Mexico	7,201	31,484	30	36	7,173	31,448
Southern California	988	5,002	6	12	982	4,990
South Atlantic	400	3,157	2	5	398	3,152
Northern California	468	2,706	4	6	464	2,700
Eastern Gulf of Mexico	180	2,397	3	6	177	2,391
Navarin Basin	a	2,054	a	16	a	2,038
Central California	240	1,587	2	4	238	1,583
Mid-Atlantic	90	897	1	2	89	895
St. George Basin	a	754	a	4	a	750
Washington-Oregon	130	486	a	1	130	485
Beaufort Sea	a	682	a	4	a	678
North Atlantic	17	245	1	1	16	244
Straits of Florida	a	55	a	a	a	55
Chukchi Sea	a	600	a	3	a	597
Gulf of Alaska	a	36	a	1	a	35
North Aleutian Basin	a	26	a	a	a	26
Norton Basin	a	34	a	1	a	33
Kodiak	b	b	b	b	b	b
Hope Basin	b	b	b	b	b	b
Shumagin	b	b	b	b	b	b
Cook Inlet	b	b	b	b	b	b

[a] Estimated to be less than 0.5 million $1987.
[b] Resources for these areas are estimated to be negligible; thus no production is expected, and social costs are estimated to be less than 0.5 million $1987.
Source: Minerals Management Service, *Five-Year Leasing Program,* 1987, p. 79. Low price corresponds to $15.75 per barrel starting price in 1987; high price to $32.50 per barrel.

in a statistically significant causal variable such that the expected number of sales increases by one. This estimated change in a causal variable is then compared heuristically to the changes that were likely to result from analysis.

The regression results are presented in Table 6-9. The number in each row that is not in parentheses is the estimate of the coefficient for the variable named in the left-

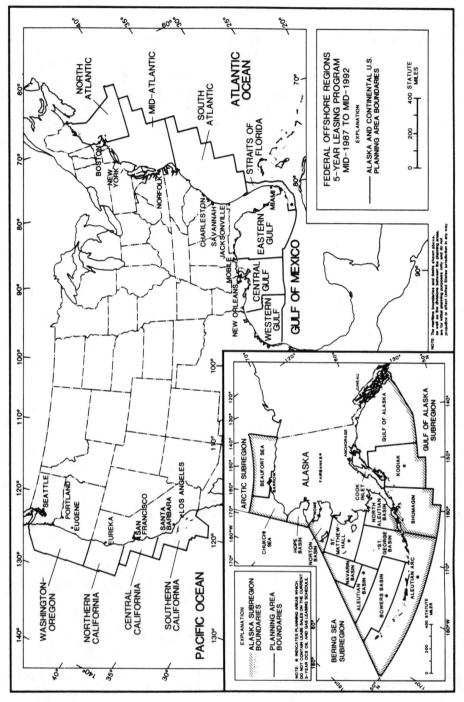

Figure 6-3. OCS Planning Areas (*Source*: Minerals Management Service, "Oil and Gas Leasing/Production Program: Annual Report/FY 1988," OCS Report, MMS 89-0055.)

Table 6-9
Poisson Regression Results[a]

Variable Sample	(1) 26	(2) 26	(3) 22	(4) 20
Constant	.06	.85	.25	−1.01[b]
	(.31)[c]	(1.38)	(1.24)	(−2.35)
Net social value	.50E-.04[b]	.25E-04	.44E-04[b]	.0006[b]
	(4.18)	(1.32)	(3.61)	(2.05)
Industry rank	—	−.08[b]	—	—
	—	(−2.27)	—	—
Comments	—	−.15E-03	—	—
	—	(−.66)	—	—
Litigation	—	.87	—	—
	—	(.63)	—	—
Log likelihood fn.	−51.42	−49.95	−51.37	−10.26
		Policy Significance		
dx|dsales = 1	16 billion	10 places	—	1.3Bil.[d]

[a] Dependent variable is the log of sales.
[b] Indicates asymptotic significance at the 95% level.
[c] t statistics in parentheses.
[d] Change required to increase probability of two sales to .5.

hand column. The first two columns are based on a sample that includes all 26 planning areas. Column three reports results from altering the sample assuming different decision processes for four early decision areas in Alaska where sales were set to zero by the Department of the Interior prior to collecting all the data. Column four deletes the two major producing areas in the Gulf of Mexico as well.

Column one of Table 6-9 reports the answer to the statistical significance of the analytic inputs for the full sample. Net social value is a significant determinant of the number of sales.[24] Results not reported are similar for industry rank and the groups specified by the Department of the Interior. However, the issue of policy significance remains. Given the size of the coefficient, a critic might ask, "How large a change in the net social value is necessary to change the expected number of sales by one?" This is particularly relevant to the influence of net social cost measures on the expected number of sales.

The estimates reported here imply that a $16-billion change in the net social value of an area is necessary to change the expected number of sales by one. Since only two areas (the central and western Gulf of Mexico) are estimated to have a net social value exceeding $5 billion and the largest social cost in any area was less than $50 million, the expected number of sales is unaffected by changes in net social value that are likely to be observed.

The second column of Table 6-9 reports regression results that incorporate other variables associated with the statutory balancing criteria for determining sales as well as the probability of litigation (a separate item that might be an element of the secretary's decision criteria). The only statistically significant regressor in this case is industry rank. The size of the coefficient indicates that the change of rank must be very large (approximately 10 out of 26) in order to change the expected number of sales by one.

The results for the entire sample in Table 6-9 indicate that whereas there is some evidence for the statistical significance of value measures in the choice of the expected number of sales, the value of the coefficients indicate that there is no evidence for policy significance. However, a different decision process may have been applied to different subsamples of the areas and are investigated below.

Estimates from varying the sample to exclude the Alaskan planning areas that were assigned zero sales early in the planning process are presented in column three. The results indicate again that the value measure can be statistically significant even though it is policy insignificant since the coefficient is smaller than in the previous table.

Finally, column four reports the results when the central and western Gulf of Mexico planning areas are also dropped from the sample. These two areas are well-established producers and have net social value measures more than six times greater than any other planning area. It is possible that a different decision process was applied to these areas to the extent that there was little uncertainty about the number of sales to be held; each area has five sales scheduled.

The resulting sample reported in column four contains the 20 planning areas that represent the major areas of controversy—those areas where analysis may have the most to contribute. However, all 20 planning areas in this sample have one or two sales indicating that the Poisson distribution is probably no longer appropriate. For this sample, a binary choice model is more appropriate. A probit analysis—which uses the normal distribution to estimate the probability of an event—was used to test the significance of the regressors for predicting the chance that two sales instead of one would be held.[25] The results of estimating the probit model for the areas with one and two sales are presented in column four. The results indicate that net social value is a significant determinant of the probability that an area will have one or two sales.

The question of policy significance is difficult to address in the probit model since the predictions are probabilities of observing one or two sales. The measure used to determine policy significance is the change in the regressor necessary to change the probability of observing two sales to .5. This change is somewhat arbitrary as it only indicates the change necessary to yield an even chance of observing one or two sales. Policy insignificance is again indicated in that the net social value of the average area with one sale would have to increase by 1.3 billion—two and one half times the mean value of 540 million—just to increase the probability of a second sale to .5.[26]

Judicial review of the planning process for OCS lease sales has given the Department of the Interior an unusual freedom to use cost benefit analysis as a major factor in balancing competing congressional objectives. The results reported here indicate that whereas there is a statistical relationship between cost benefit measures and the expected number of sales, the net social value measure must change by an impracticably large number to cause a change in the expected number of sales or in the probability of another sale for a given planning area.

In general, one cannot reject the hypothesis that analysis mattered within the broad confines of the many balancing criteria available to the secretary. It appears, however, that less analysis could have mattered just as much as the complex net social value measures in the choice of the number of sales. One does not need to look far for hypotheses about the cause of this behavior. As William Bettenberg, former director of the Minerals Management Service described the expectations of the Department of the Interior:[27]

When we were fairly fresh into the process of evolving the next five-year program, we had a new undersecretary (of the Interior). I briefed her, and went through this entire process. I got to the point where we adopted the five-year program, and she said, 'Well, then what happens?' and I said, 'Well then we go to court,' and she was taken aback. I said, 'Well, the Andrus program was litigated, the Watt program was litigated. We know that this will be litigated.'[28] We have to take that into account about how we go about the process, how we write every document.

The planning process and supporting analyses serve several purposes in the determination of the pace of leasing of which informing the secretary is only a part of predictable strategic planning for the legal battle to follow.

Part Three

Emerging Policy Issues

The first two sections of this book focus on existing practice and past and ongoing issues in the management of OCS lands. However, management practices evolve constantly. No issue is ever completely resolved and new issues emerge that create demands for new management practices and new policy analyses. These two concluding chapters present some emerging issues in the management of the OCS and draw together the research and policy recommendations of earlier chapters.

The methods for managing nonfuel minerals are one emerging issue of OCS management. Many nonfuel minerals from the OCS are forecast to become important economically, although there is only minor administrative experience and little organizational structure in place. Chapter 7, authored by James Broadus and Porter Hoagland, surveys the existing institutions, the potential resources, and the emerging issues in the management of OCS nonfuel minerals.

In Chapter 8, two OCS managers, William Bettenberg, former director of the Minerals Management Service, and Carolita Kallaur, deputy associate director of the Minerals Management Service, present their views of the many remaining areas of uncertainty about OCS management. These interviews also serve as a reminder that it is people, and not a faceless system, who manage the OCS. In these interviews and in the recommendations for research and for policy that follow a variety of emerging issues in OCS management are discussed. These issues include the long-term effects of low level pollution and the roles of risk perception and risk communication.

Chapter 7

Nonfuel Minerals†

Oil and gas resources are not the only mineral assets of the OCS. An abundant variety of nonfuel mineral deposits have also been identified. Examples include sand and gravel in the New York bight, Beaufort Sea, and in areas offshore of California and Hawaii, placer deposits containing chrome, gold, platinum, titanium, and other heavy minerals, phosphorites along the southeastern U.S. margin and the coast of Southern California, manganese nodules on the Blake Plateau, cobalt crusts in areas around Hawaii, and marine polymetallic sulfides (MPS) in the Gorda Ridge area. Although these OCS nonfuel minerals are currently of only minor economic importance (as compared to OCS oil and gas resources), interest in them has grown markedly in recent years, and efforts have begun to clarify procedures for their exploration and development. Like their hydrocarbon counterparts, OCS nonfuel minerals are managed by the Minerals Management Service. However, most OCS nonfuel minerals are not commercially exploitable at present.

Nonfuel minerals on the OCS remain unworked except for the limited development and production of sulfur and associated salt deposits in the Gulf of Mexico. Due to geologic and economic uncertainty, the primary policy issue with respect to these minerals is one of generating information and allocating discovery and exploration effort between public and private interests.

Worldwide, the contribution of marine nonfuel minerals to minerals supply is very small compared with the more conventional onshore sources of the same commodities.[1] In the United States, OCS nonfuel deposits eventually might provide additional resources for at least 26 materials (although the magnitude and timing of these additions still require serious study). For most minerals, a period of exclusive production from successively costlier onshore deposits can be expected until a cost level is reached at which the least-cost marine deposits—as with hydrocarbons or sulfur—join into total production. Beyond that point, the respective shares of total output coming from onshore and marine sources would depend on how much additional output could be obtained from each source at incrementally higher cost levels.

Increasing attention has been paid to OCS nonfuel minerals, as evidenced by recent congressional hearings and studies of marine nonfuel minerals policy.[2] Given the cur-

---

†This chapter is by James M. Broadus and Porter Hoagland III.

rent economic potential of marine nonfuel minerals, the reasons for this attention are not immediately apparent. However, the discovery and growing understanding of new forms of ocean minerals, such as marine sulfides, certainly fuels interest and reinforces an optimistic outlook for long-run minerals supply.

STRATEGIC MATERIALS

The availability of certain nonfuel minerals (designated "strategic" or "critical") as a supplemental source of metal commodities has been an important national policy goal to past federal administrations. In 1983 the Reagan administration reemphasized this goal, with particular focus on the potential resources of some of these minerals in the ocean. When President Reagan proclaimed an Exclusive Economic Zone (EEZ) for the United States, he said that "recently discovered deposits there [in the Exclusive Economic Zone] could be an important future source of strategic minerals."[3] Concerns for strategic minerals are reflected also in the name of the Minerals Management Service office, the Office of Strategic and International Minerals, which promotes the development of marine nonfuel minerals on the Outer Continental Shelf and within the EEZ.

Much disagreement exists about exactly what materials are or are not "strategic." A recent careful attempt to analyze this issue narrowed a long list to four "first tier" commodities: chromium, cobalt, manganese, and the platinum group metals. All four are potential OCS nonfuel minerals. The primary interest groups are government agencies with responsibility for supplying the national defense structure in times of crisis.

As a component of national minerals and materials policy, the management of OCS nonfuel minerals must be considered in the context of other factors that condition mineral supply. If there is some benefit to reducing the risk of economic disruption from variations in supply, the costs of encouraging marine mineral development should be compared to the costs of maintaining stockpiles, encouraging onshore mineral development, sponsoring basic and applied research into substitute minerals, recycling, and mining beneficiation technologies, and conservation. In 1985 the U.S. Office of Technology Assessment concluded that options based on substitution, conservation, or production from alternative conventional sources are superior to marine mining as approaches to reduced import dependency.[4]

ORGANIZATIONS AND JURISDICTIONS

Under the general authority of several broad policy mandates, the Department of the Interior is responsible for encouraging private research and development in domestic mining, metallurgy, and critical materials, both onshore and offshore. Under the specific authority of OCSLA section 8(k), the Department has promoted OCS nonfuel mineral development since an early sale of OCS lands for phosphorite minerals off the coast of Southern California in 1961. When the oil platform blowout occurred in the Santa Barbara Channel in 1969, public concerns about the external effects of industrial activity in the oceans helped to delay plans for leasing OCS lands for nonfuel minerals development.

However, in 1974 the Department's Bureau of Land Management (BLM) published a draft environmental impact statement with details on the proposed disposal of OCS lands for phosphorite and sand and gravel resources. Public reaction to the draft was mostly negative, and the BLM postponed its effort.

In 1975 the Department established an Ocean Mining Administration (OMA) to coordinate its ocean mining efforts, including marine geological research activities conducted by the Department's U.S. Geological Survey (USGS) and studies on the environmental effects of manganese nodule metallurgical processing by the Department's Bureau of Mines (BOM). The Ocean Mining Administration was short-lived. It spearheaded the Department's support of domestic deep seabed legislation in the late 1970s, until the Deep Seabed Hard Mineral Resources Act of 1980 gave primary management authority over deep seabed minerals (specifically manganese nodules) to the National Oceanic and Atmospheric Administration (NOAA) in the Department of Commerce. In 1979 the USGS published a Program Feasibility Document on a proposed OCS nonfuel minerals program recommending a prototype lease sale. This recommendation went largely unheeded for three years during a change in administration.

Existing Authority: OCS Lands Act

In 1983 the Department of the Interior created the Office of Strategic and International Materials (OSIM) within the Minerals Management Service. OSIM was directed to proceed with activities leading to the disposal of OCS lands for nonfuel mineral development. Based upon the guidance of an interagency task force, the Minerals Management Service began to construct a regulatory regime to carry out the provisions of Section 8(k) of the OCS Lands Act. The agency planned three tiers of regulations directed at prelease exploration, leasing, and postlease activities.

As part of its new responsibility, OSIM implemented an innovative concept known as the federal-state "task force" (used earlier in the leasing of public lands onshore for oil shale and geothermal resource development). These task forces have been aimed at bringing coastal state managers into the leasing process at an early stage. The hope of the Minerals Management Service is that this process might reduce later delays associated with the intergovernmental aspects of the OCS disposal process. The specific activities of each task force have varied from case to case. Thus far, the task forces have been funded to conduct a range of activities from economic feasibility studies to oceanographic research, to public hearings, and to the drafting of environmental impact statements. The states of Alabama, Alaska, California, Georgia, Hawaii, Louisiana, Mississippi, North Carolina, Oregon, Texas, and Washington (as an observer) already participate as members of the task forces.

Legislative and Regulatory Initiatives

The enactment of the OCS Lands Act in 1953 was an exercise of the constitutional authority of Congress to dispose of public lands. (Although OCS lands are not true "public lands," Congress defined them in the act to be subject to U.S. jurisdiction, control, and power of disposition as discussed in Chapter 1.) The brief, nonspecific paragraph of the act that relates to the disposal of OCS lands for nonfuel mineral development leaves a considerable amount of discretion to the secretary of the Interior to set the terms and conditions of an OCS nonfuel mineral lease. One of the only specific requirements is that leases must be sold on the basis of competitive cash bonus bidding at an auction. Not surprisingly, industrial interests have been opposed to this disposal method. The Minerals Management Service reportedly has considered modifications to the traditional bonus bidding method utilized for the fossil fuels, but these considerations have not yet been widely published.

In large part because of the efforts at the OSIM to construct a regulatory system for nonfuels that would dispose of OCS lands competitively, an alternative legislative initiative has been pursued. In 1986 and 1987 identical bills were introduced in successive sessions of the Congress to change the disposal method found in the OCS Lands Act to an exploration licensing system.[5] The proposed system would include a "preference right" to a development-production permit in the event that a deposit of commercial potential is discovered. The current legislative proposal has been called the National Seabed Hard Minerals Act. In a general sense, this proposal is a throwback to one of the earliest methods of mineral disposal—the location-patent system of the 1872 Mining Law used on the federal public lands of the western United States. Significant differences can be found in the current proposal, of course, including the right to an exclusive exploration license and the possibility of a royalty charge upon production. The proposed National Seabed Hard Minerals Act has been drafted much more closely along the lines of the Deep Seabed Hard Mineral Resources Act (the legislation governing the allocation of entitlements to manganese nodules beyond national jurisdiction).

Multiple Agency Management

Jurisdictional ambiguities (geographic and regulatory) are a common feature of the management of OCS nonfuel minerals. This fact is well recognized by participants in the process but seldom has been subject to serious academic analysis. Such ambiguities can be expected in any situation when one agency has general management responsibility where the resources are found, but another agency manages the specific resources that might be extracted.

In the case of nonfuel minerals found on the OCS and on the deep seabed, at least two federal agencies, the Minerals Management Service and the National Oceanic and Atmospheric Administration, have statutory authority to regulate exploration and exploitation. Although there is an apparent clear division of responsibility at the continental shelf "boundary," the actual division is in fact uncertain. The creation of the Exclusive Economic Zone, which extends 200 nautical miles from the coastal "baseline," further clouds the jurisdictional issue. The Solicitor's Office of the Department of the Interior has published an opinion that concludes that the Interior Department has the authority to manage nonliving resources of the EEZ in areas that extend beyond what might be considered as the geographical continental shelf.[6] Yet, there has been no official delimitation of the OCS.

Because of the preliminary state of knowledge about most OCS nonfuel minerals, a major share of management activity in the nearterm will necessarily involve programs of scientific research. Management ambiguities may be difficult to unravel for the many promotional, research and development, and data collecting and handling responsibilities of the Commerce and Interior departments. Agency roles and responsibilities overlap significantly in this area. NOAA and the USGS have recognized the need to co-ordinate individual and overlapping agency responsibilities for research in the water column, beneath the ocean floor, and on the ocean. Moreover, these two agencies have established a liaison office and have agreed to complement each other's activities in a bathymetric survey of the EEZ. In spite of such advances in mutual understanding, the Commerce and Interior departments may continue to share jurisdiction. Furthermore, preliminary indications show that, in several cases, those private firms that are already dealing with the National Oceanic and Atmospheric Administration on manganese nodule development are the same firms that have shown interest in potential OCS nonfuel resources.

Where areas containing potentially valuable natural resources are subject to multiple agency management and associated ambiguities, the pace of resource development can be subject to two offsetting effects. Multiple systems of rules covering a single resource or activity can impose additional costs that postpone the time when a resource will be explored and developed. This is especially true in a case such as marine nonfuel minerals where significant uncertainties are present about the physical operating environment. Private firms can be reluctant to commit investments in exploration (much less to establish long-term development plans in an uncertain legal environment). On the other hand, if the resource can be independently explored and developed by separate parties, each may accelerate its own development activities in order to discover and recover as much as possible before their competitor (the well-known "common pool" effect).

Under certain conditions, however, some jurisdictional overlap and managerial rivalry can be beneficial, enhancing both the quality of policymaking and the flow of information to interest groups such as private firms, environmental groups, and states. Especially in the early stages of an evolving legal regime, private interests who are contemplating investment in resource development may find enhanced access and greater range of influence on agency decisions when more than one agency is centrally involved in the process. Where the allocation of public funds and the selection of research projects depend critically (as they do for marine hard minerals) on scant scientific knowledge, it is especially important that resource managers have both up-to-date scientific information and balanced appreciation of its significance. Multiple agency management responsibility might enhance flexibility in the face of uncertainty and provide increased scope for combinations of agency strengths and specialties. Because a complex (and still poorly defined) variety of functions will be necessary to convert OCS nonfuel resource potential into a flow of economic supplies, premature monopolization of all these functions by a single agency could sacrifice the benefits of multiple agency management at a time when they seem most essential.

DESIGNING AN OFFSHORE NONFUEL MINERAL SYSTEM

All minerals have geologic and end-use characteristics that are distinguishing features. Several government studies and laws have recognized these features and, based on observed differences, have recommended the need to manage nonfuel minerals in a different manner than hydrocarbon minerals. It has been stated frequently by those engaged in the policy debate over management systems for offshore nonfuel minerals that differences in industrial structure and technology between the hard minerals industry and the oil and gas industry require an entirely different set of offshore access provisions. This concept was incorporated in the proposed National Seabed Hard Minerals bill virtually without discussion and certainly without adequate elaboration. The differences most often referred to are:

1. Investors face more uncertainty and greater risk in marine nonfuel minerals exploration.
2. Technology for marine hard minerals exploration and development is less developed than for marine hydrocarbons.
3. The oil industry has more experience in searching for minerals beneath the surface, and exploration success in marine hydrocarbons can be converted more readily into production.
4. Marine hard minerals may require more extensive drilling or testing and a longer period of development without benefit of revenues than for hydrocarbons.
5. The hard mineral mining industry has fewer financial reserves than the oil industry and is unable to pay large, front-end cash bonuses.

The implication drawn from these differences by prospective ocean miners is that a competitive bidding system for access and development entitlements (particularly one based on up-front payments such as the OCS Lands Act) is inappropriate for marine hard minerals. Prospective ocean miners prefer the licensing system found in the Deep Seabed Hard Mineral Resources Act because it collects only a minor tax on production. The provisions described in the proposed National Seabed Hard Minerals Act promise somewhat greater financial consideration to the public in exchange for the right to explore.

An important policy issue associated with noncompetitive licensing is that licenses are not necessarily distributed to the most efficient explorer or developer. Instead, they are issued to the first in line at the "land office." If licenses can be assigned by one firm to another, then, except for additional negotiation costs, this poses no problem for economic efficiency.

Whereas noncompetitive licensing is capable of achieving economic efficiency, the public would not receive a financial return. However, there appears to be no tangible reason why licenses for OCS nonfuel minerals could not be issued competitively (as they must under OCSLA 8(k)) so that access is allocated to the most efficient producer. Due to the high degree of geologic uncertainty and the current low level of industrial interest, however, expected economic rents could be low or even nonexistent. And thus bids would also be low or nonexistent. To compensate for the lack of rents, methods other than bonus bidding—such as profit share bidding or work commitment bidding, among others—that allocate access competitively have been suggested. In general, these methods of access impose enforcement and other administrative costs that are not encountered in a license-permit system. Moreover, there may not be enough commercial interest to hold a competition for access.

From the perspective of the resource manager who expects rapid technological change and uncertainty, several guidelines seem important for allocating the OCS nonfuel mineral lands. Achieving congressional objectives for nonfuel minerals may require maintaining a high degree of adaptability, diversifying to avoid commitment to a single outcome. However, as knowledge of OCS nonfuel minerals grows, and if the mineral resource potential proves sufficient to generate a stronger market for access, then the nonfuel OCS management may become more similar to that for oil and gas.

In the nearterm it may be beneficial for the resource manager to have the authority to make small changes in a disposal system to adjust to variations in economic conditions over time as well as new emphases in public goals. This kind of adaptability might be implemented at different levels and in alternative ways such as:

1. Individual access. The adjustment of terms and conditions of property rights on a case-by-case basis. Although this could incur substantial administrative costs, the small number of expected leases in the near future suggests that such a system might be administratively feasible.

2. Dual system. Multiple disposal methods might be employed for the same minerals across space or time, based upon the model found in the 1920 Minerals Leasing Act. This Act established a dual system in which solid minerals (e.g., phosphates, sulfur, salt) are leased competitively in "known geological structures" but are leased on a first-come, preference-right basis in areas where geological structures are unknown. The most recent version of the National Seabed Hard Minerals Resources Act, H.R. 2440, contains language that would allow the establishment of a dual system.

3. One system with marginal adjustment. One disposal method might be employed with marginal adjustments over time (as exemplified by the oil and gas provisions in the OCS Lands Act). Terms and conditions could be modified from time to time on future disposals to respond to changing market conditions.

4. Interim system with a sunset clause. This would entail a system like the proposed National Seabed Hard Minerals Act that would expire after a number of years (or perhaps after one round of licensing). Faced with virtually the same issues for nonfuel minerals on the U.S. public lands, the Public Land Law Review Commission recommended an "interim system" in 1970 and placed a premium on adaptability.

An important point is that adaptability is not equivalent to the exercise of discretion over access already granted through a lease or some other agreement. Adaptability can be incorporated into a disposal system for OCS nonfuel minerals without an increase in managerial discretion and its associated uncertainties.[7] Adaptability involves the adjustment of basic disposal methods such that the probability of achieving policy goals through future OCS land disposals is increased. The limits of adaptability may soon be tested for the nonfuels case under the provisions of the OCS Lands Act. It appears that the act could be more adaptable for the oil and gas minerals than for the nonfuels. Likewise, until the 1989 version as H.R. 2440, the legislative proposal (National Seabed Hard Minerals Act) did not appear to be any more adaptable than the OCS Lands Act in the sense described here.

POTENTIAL OCS NONFUEL MINERAL RESOURCES

Comparative Costs

In the placid, shallow waters of protected bays or estuaries that typically are under the control of the states, dredging costs for loose materials, such as sand and gravel, placer minerals, or phosphate, may be comparable to those onshore. Indeed, this is little more than an extension of conventional onshore production. For more exposed, high energy (weather and waves) offshore environments, much greater costs can be expected. Mining costs tend to be case-specific, but industry sources suggest that seabed dredging for these materials would cost three–five times more than inland dredging.

When a mining technology is more costly than another for a given level of ore, it still can be competitive if the ore grade is rich enough to compensate with higher metal yield or if the deposit is large enough to spread fixed costs over greater levels of output. For example, although average offshore drilling and equipping costs tend to be three–four times larger than onshore costs for oil and gas, the very large size of producing offshore deposits allows them to be competitive. Similarly, other marine deposits would have to offer compensating grade or size premiums to be competitive. Under some local conditions with locational or deposit-size advantages in delivered cost, offshore sand and gravel materials overcome the usual cost differential.

Mineral Types, Reserves, and Resources

From a general standpoint, OCS nonfuel mineral prospects can be classified into shallow coastal and deepsea deposits. Table 7-1 presents descriptive statistics comparing onshore and offshore production, estimated revenues, and resource estimates for nonfuel deposits with potential marine sources.

Shallow coastal deposits are the first general class of marine nonfuel minerals. These deposits are generally found in waters less than 200 meters and include the following six mineral types: sulfur, sand and gravel, calcium carbonate, marine placers, phosphorite deposits, and lode minerals.

1. *Sulfur* has been recovered commercially off the coast of Louisiana since the early 1960s. Sulfur is used as a chemical reagent and in the production of fertilizer. Salt-capped sulfur domes are mined using the Frasch process of injecting hot water to melt the sulfur and air pressure to force the melt to the surface. Salt is recovered nearby and used to make a brine solution that acts as a drilling fluid in sulfur extraction. Because salt is also an OCS nonfuel mineral, the rights to salt deposits are often sold together with sulfur rights. Offshore sulfur production is now limited to a single operation, producing approximately $36 million in revenues annually. The recovery of waste sulfur from pollution control equipment may replace Frasch sulfur entirely by the year 2000. Since 1953 only six lease sales have been held for sulfur minerals on the OCS (Table 7-2). The most recent sale, held in February 1988, attracted $15 million in high bonus bids on 14 tracts in the Gulf.

2. *Sand and gravel* are produced within the jurisdiction of coastal states by small dredging operations for construction aggregate. For decades, deposits of these

Table 7-1

Descriptive Statistics: Offshore Nonfuel Minerals

U.S. Marine deposits	Material commodity	(A) U.S. Marine production[a] (MT×10^3)	(B) U.S. Mine production (MT×10^3)	(C) Estimated average price ($/MT)	(D) (A)×(C) Marine revenues ($×$10^6$)	(E) (B)×(C) U.S. Revenues ($×$10^6$)	(F) (D)×100/(E) Marine share of U.S. revenues (%)	(G) U.S. Marine reported speculative resources (MT×10^3)	(H) U.S. Identified onshore resources (MT×10^3)	(I) (G)×100/(H) U.S. Speculative marine resources compared to U.S. identified resources[b] (%)
Sand and gravel	Sand and gravel	3,000	865,900	3	9	2,598	1	665,778	65,000,000	1
Shell	Calcium carbonate	14,000	(1,400,000)	6	84	(8,400)	1	large	very large	small
Sulfur	Sulfur	381	11,200	105	40	1,176	3	27,125	1,000,000	3
	Barite	—	343	31	—	11	—	2,087	90,720	2
	Phosphate rock	—	40,000	24	—	960	—	4,615,000	9,250,000	50
Mineral placers	Rutile	—	—	364	—	—	—	12,156	7,801	156
	Ilmenite	—	—	49	—	—	—	180,537	87,091	207
	Titanium	—	16	12,236	—	196	—	—	—See: Rutile, Ilmenite—	—
	Zirconium	—	—	182	—	—	—	25,039	12,207	205
	Hafnium	—	—	231,483	—	—	—	250	127	197
	Yttrium	—	—	35,020	—	—	—	3,450	28,123	12
	Thorium	—	1	35,850	—	1	—	—	—	—
	Chromite	—	0	42	—	0	—	—	—	—
	Gold	—	1	10,600,000	—	800	—	30,158	9,979	302
	Platinum	—	—	9,000,000	—	—	—	1	8	10
Crusts	Platinum	—	—	9,000,000	—	—	—	1	9	1
	Cobalt	—	0	25,353	—	0	—	42,267	1,270	3328
	Nickel	—	1	5,026	—	5	—	22,684	13,880	163
	Manganese	—	0	141	—	0	—	1,128,751	66,770[c]	1691
Massive sulfides	Copper	—	1,170	1,475	—	1,726	—	—	382,000	—
	Zinc	—	210	893	—	188	—	—	65,000	—

a U.S. marine production occurs predominantly in the territorial sea.

b Reported grades may be incomparable (particularly in the case of estimated crust resources).

c Very low grade manganese, less than 20 percent wt.

Source: P. Hoagland III and J. Broadus, "Seabed Material Commodity and Resource Summaries," Woods Hole Oceanographic Institution, WHOI-87-43, 1987.

Table 7-2
Gulf of Mexico OCS Sulphur and Salt Lease Sales

Year	Tracts bid on	Tracts leased	Total bonus (million $)
1954	5	5	1.2
1960	1	1	.1
1965	50	50	33.7
1967	8	1	.0
1969	38	4	.7
1988	14	14	15.1

Source: Minerals Management Service.

materials have been mined for their use as construction aggregate or for beach nourishment. In countries such as England, the Netherlands, and Japan, marine sources make a major contribution to total sand and gravel supplies. In the United States, these sources still account for only about 1 percent of total production, but the potential is large.

In 1987 the U.S. Bureau of Mines completed a study commissioned by the Minerals Management Service of the prospects for offshore production of sand and gravel in the United States.[8] The study suggested that "significant potential" may exist for the development of sand and gravel off New York and Boston in the nearterm. At present, production is accomplished by pumping or dredging, often tens of kilometers offshore. Practical recovery depths are less than about 50 meters. High transport costs limit these construction minerals to local market areas, with great geographic variety in price among markets.

In 1983 the Minerals Management Service initiated steps toward the leasing of OCS lands for sand and gravel resources off Alaska. The oil and gas industry had expressed an interest in these deposits because of their potential use as a material for the support of production platforms in the icy Arctic region. However, due to slumping oil prices, industry interest in Arctic production waned, and the sand and gravel sale was cancelled.

3. *Calcium carbonate* is recovered primarily in the form of the mineral "aragonite" and is used in cement, glassmaking, and foundry applications. (Shell, mentioned above with other aggregates, is also a form of calcium carbonate.) In 1986 the Minerals Management Service issued a prelease prospecting (geological and geophysical) permit for "carbonate sands" on the OCS lands off the Florida Keys. Calcium carbonate is mined onshore in the form of limestone (of which vast deposits exist). Limestone is used as a construction material (as either "crushed" or "dimension" stone) and also to produce lime for steelmaking, water purification, and pollution control. The U.S. Bureau of Mines classifies some organically generated forms of calcium carbonate (such as precious coral and pearl) as types of gemstone. Exploitation of these organically generated mineral forms are considered a fishery and a precious coral fishery exists off the coast of Hawaii.

4. *Marine placers* include deposits of light-heavy minerals such as chromite or titanium oxides (ilmenite, rutile, leucoxene) and other "associated" minerals (zirconium, monazite containing yttrium, thorium, and other rare earths) and heavy-

heavy minerals like native forms of precious metals or tin. Again, with the sponsorship of the MMS, the Bureau of Mines conducted engineering and cost studies of a chromite placer off the coast of Oregon, titanium placers off the coasts of Virginia and Georgia, and a gold placer off the coast of Alaska.[9] Only the gold placer is being worked (although there has been some recent prospecting efforts on the others). Off Nome, Inspiration Mining Company is presently recovering gold from an offshore placer deposit within the Alaskan territorial sea. A small portion of this deposit may extend beyond Alaskan jurisdiction onto OCS lands. Marine placer minerals so far reported do not seem to exhibit a much larger size or higher grade than their onshore rivals, and Emery and Noakes have shown that strong physical constraints generally will limit the distribution, grade, and accessibility of marine placers.[10] Table 7-3 compares two estimates of the costs for developing and producing the Nome deposit.

5. *Phosphorite* deposits are found off the coast of Southern California, North Carolina, and Florida. The potential for phosphorite development off the U.S. Atlantic Coast near North Carolina has generated recent interest, and a jointly sponsored state and federal effort has been directed at examining the commercial potential

Table 7-3
Nome Gold Placer Costs

	Bureau of Mines (January 1987)	OTA
Deposit kind	Gold Placer	Gold Placer
Grade	0.6 gram per yard^3	0.35 to 0.45 gram per yard^3
Size	35,000,000 yard^3	80,000,000 yard^3
Distance to shore unloading point	0.5 to 5 miles	0.5 to 10 miles
Maximum dredging depth....	80 feet	90 feet
Annual mining capacity— tonnage dredged	1,632,000 yd^3	4,500,000 yd^3
Mining system	Used seagoing bucket line dredge with full gravity processing	Used seagoing bucket line dredge with full gravity processing
Mining system operating days	150	150
Shore processing plant	Minimal for final cleaning of gold concentrates	Minimal for final cleaning of gold concentrates
Capital costs (million $)		
Dredge		5
Plant and other		10–15
Total...................	$9.1	15–20
Direct cash operating costs $US per yard mined	2.00	1.55
Comments (OTA'S)	Technically feasible and appears economically profitable	Technically feasible and appears economically profitable

Source: Office of Technology Assessment, *Marine Minerals,* 1987.

of deposits there. The only lease sale held to date for OCS nonfuel minerals other than sulfur/salt has been for phosphorite deposits off the coast of Southern California in 1960. These leases were relinquished shortly after the sale ostensibly because of the existence of unexploded naval ordnance at the site. Some interest has been shown in the development of new borehole mining methods that may enhance access to subseabed phosphates, but this also favors expansion of deep rival resources onshore.

6. *Lode minerals* include deposits such as the barite of Castle Island, Alaska. Barite, or barium sulfate, is used as a weighting agent in oil and gas drilling fluids. Until 1980 barite was produced from Castle Island within the waters of the state of Alaska. The Castle Island mine—a particularly rich deposit—was mined out above sea level from 1966 to 1969. Subsequent exploration work by the Inlet Oil Company discovered subsea reserves of approximately 2 million metric tons, and the mine was continued below sea level, effectively turning it into a marine deposit. The need for barite is tied closely to drilling activity in the oil and gas industry and the Castle Island mine is not currently in production. Lode deposits of the OCS lands are unknown and will probably be difficult to discover.

Deepsea deposits are the second general class of nonfuel minerals, including cobalt-enriched ferromanganese crusts on the flanks of seamounts between 1,000–4,000 meters in depth, marine sulfides precipitated around hydrothermal vents at crustal spreading centers found between 2,000–2,500 meters, and certain deposits of marine phosphorites found on seamounts between 1,000–4,000 meters. At this stage, attempts to characterize potential mining costs for marine sulfides and cobalt crusts are speculative at best. No technologies are known for breaking, sorting, and lifting these hard-rock deposits at such great depths, and only the most preliminary mining concepts have so far been presented. No method is known by which the crusts can even be selectively extracted in quantity without also extracting much barren substrate material, and practically nothing is known about the thickness (size) of the sulfide deposits. Development of techniques, such as a hard-substrate drill, to overcome these shortcomings is a priority in seabed minerals exploration.

"Resource" estimates have been reported for cobalt crusts in certain areas of the central Pacific, but these are based largely on very limited sampling or hypothetical grade-concentration combinations.[11] Cobalt crusts occur on the OCS near Hawaii and other Pacific territories of the United States. They also occur on the current-swept Blake Plateau off the Eastern Coast of the United States.

The highest quality (grade and density) manganese nodules are located beyond U.S. jurisdiction between the Clarion and Clipperton fracture zones on the Pacific seabed. Several U.S. firms hold exploration licenses in this area, issued under the authorization of the Deep Seabed Hard Minerals Resources Act. Lower quality nodules are found within the U.S. EEZ. High densities of nodules (but of low quality) are found on the Blake Plateau off of the southern Atlantic Coast of the U.S. These lower quality nodules are merely mineral "occurrences" whose economic significance is minor at best.

For marine sulfide deposits, geological and geochemical inference provide virtually the only basis for estimates of potential quantities in-place because of the extremely limited number of observations (approximately 50 sites have been sampled to date) and the absence of data on deposit thickness. Some speculative extrapolations of minerals in vent deposits on the midocean ridge based on geochemical deposition models have been attempted.[12] In 1983 the Minerals Management Service considered the disposal of lands for purported marine sulfide resources on the OCS near the Gorda Ridge (an active seafloor spreading center off the coasts of Washington and Oregon). A draft environmental impact statement for the Gorda Ridge received negative comments particularly because of its lack of resource information (sulfide deposits were not discovered on the ridge until 1986). To date, marine sulfides have been discovered on the OCS only at the Gorda Ridge.

Exploration and Technological Advance

For most OCS nonfuel minerals, traditional sampling and dredging technologies would probably be used to explore for and recover the minerals. Basic search strategies for OCS nonfuel mineral exploration are presented in Table 7-4.

During the 1960s and through the early 1980s, much effort was spent in the development of technologies for exploration and development of manganese nodules.[13] In contrast, technologies for finding and working nearshore deposits (like sand and gravel) and placers are in a relatively mature stage and have not experienced the same degree of recent technological advance. A summary of relevant dredging technologies is presented in Table 7-5.

In the long run, continuing basic and applied marine scientific research provides an "input subsidy" for potential seabed material resources. Oceanographic knowledge already has been used successfully in locating onshore occurrences of marine phosphorites, and there is some expectation that observation of marine sulfide deposits eventually will help locate commercial sites onshore. There is increasing reliance on scientific theory to target search for onshore deposits, and continuing study of marine deposits may help focus this search.

Table 7-4
Search Strategies

Approximate range to deposit	Method
10 kilometers	Long-range side-looking sonar
	Regional sediment and water sampling
1 kilometer	Gravity techniques
	Magnetic techniques
	Bathymetry
	Midrange side-looking sonar
	Seismic techniques
100 meters	Electrical techniques
	Nuclear techniques
	Short-range side-looking sonar
10 meters	Near-bottom water sampling
	Bottom images
0 meter	Coring, drilling, dredging
	Submersible applications

Sources: Adapted from P.A. Rona, "Exploration for Hydrothermal Mineral Deposits at Seafloor Spreading Centers," *Marine Mining,* 4, No. 1, 1983, 20–26.

Table 7-5
Dredging Technologies

Type	Description	Present max dredging depth	Capacity
Bucketline and bucket ladder	"Continuous" line of buckets looped around digging ladder mechanically digs out the seabed and carries excavated material to floating platform.	164 feet	Largest buckets currently made are about 1.3 yd³ and lifting rates 25 buckets per minute (1,950 yd³/hour with full buckets).
Suction	Pump creates vacuum that draws mixture of water and seabed material up the suction line.	30 feet	Restricted by the suction distance unless the pump is submerged.
Cutter head Trailing hopper	Mechanical cutters or high pressure water jets disaggregate the seabed material; suction continuously lifts to floating platform.	50–300 feet	Many possible arrangements all based on using a dredge pump; the largest dredge pumps currently made have 48″ diameter intakes and flow rates of 130 to 260 yd³/min of mixture (10 to 20% solids).
Airlifts	Suction is created by injecting air in the suction line.	10,000 feet	Airlifts are not efficient in shallow water. There may be limitations in suction line diameter when lifting large fragments.
Grab: Backhoe/dipper	Mechanical digging action and lifting to surface by a stiff arm.	100 feet	Restricted by the duration of the cycle and by the size of the bucket; currently largest buckets made are 27 yd³.
Clamshell/ dragline	Mechanical digging action and lifting to surface on flexible cables.	3,000 feet	The largest dragline buckets made are about 200 to 260 yd³/hr; power requirements and cycle time increase with depth.

Source: Office of Technology Assessment, *Marine Minerals,* 1987.

The rate of technical progress in exploration and discovery may be one area where seabed deposits are gaining on their conventional onshore rivals. Real discovery costs onshore have been rising (perhaps doubling in the past 30 years). Advances in deepsea exploration technology such as multibeam sonar, underwater photographic, and electronic imagery transmission, robotics, and deep submergence vehicles verified and refined the geophysical theories that had for several years predicted the occurrence of the hydrothermal marine sulfides deposits at oceanic crustal spreading centers. The theoretical results were largely independent of the search for commercial seabed mineral deposits as were the advances in exploration hardware. Technology developed to support offshore oil and gas operations has made a major contribution to the study of marine nonfuel minerals. Similarly, spillover benefits are also provided by investments for military and national security purposes such as the work financed by the *Glomar*

Table 7-6
Prelease Exploration Permits for Marine Minerals on the OCS

Year	Number of Permits	Permittee	Minerals Prospect	Approximate Location
1966	1	Marine Exploration	Gold Placers	Norton Sound, Alaska
1966	1	Ocean Science and Engineering	Gold Placers	Norton Sound, Alaska
1966	1	Newport News Shipbuilding and Drydock	Phosphorites	North Carolina
1967	2	Ocean Resources	Phosphorites	Southern California
1967	1	Bear Creek Mining	Phosphorites	Southern California
1969	1	Global Marine	Sand and Gravel, Heavy Minerals	New Jersey
1969	1	Ocean International	Heavy Minerals	Mid-Atlantic
1970	1	Deepsea Ventures	Manganese Nodules	Blake Plateau
1975	1	Radcliff Minerals	Sand	West Cameron, Louisiana
1986	2	DuPont	Heavy Minerals	Georgia
1986	2	Associated Minerals	Heavy Minerals	Georgia
1986	1	Technical University of Clausthal	Cobalt Crusts	Hawaii
1987	1	Inspiration Gold	Gold Placers	Norton Sound, Alaska
1987	1	Geomarex	Carbonate Sands	Florida Keys
Total	17[a]			

[a] This total does not include 32 permits issued from 1982–1987 to 7 companies in the Alaska OCS region concerning high-resolution geophysics or shallow geological investigations. These permits were directed, in part, toward sand and gravel resources that might be used for the construction of gravel islands or other support structures for offshore oil production facilities.

Sources: DOI, Program Feasibility Document: OCS Hard Minerals Leasing, 1979; R. Amato, Atlantic OCS Region; R. Kuzela, Gulf of Mexico OCS Region; D. Meyerson, Pacific OCS Region; J. Schearer, Alaska OCS Region.

Explorer submarine recovery effort of the mid-1970s and the recent navy sponsorship of the Argo/Jason system development at the Woods Hole Oceanographic Institution.

Information obtained through nonexclusive "permitted" private exploration on the OCS must be reported to the Minerals Management Service, the permitting agency. In particular, the Minerals Management Service holds all privately generated data and information obtained under nonexclusive geological and geophysical exploration permits confidential for variable 25–50-year periods. (Resource information must be made accessible to public managers upon request.) After these periods, resource information becomes available to the public. A list of permitted geological and geophysical investigations for OCS nonfuel minerals since 1960 is presented in Table 7-6. The confidential treatment of proprietary information is a sensitive issue to firms that explore on the OCS. (Recall from Chapter 3 the discussion of the billions of dollars spent on exploration and bidding as well as the conflicting effects of public versus private information.) Proposed OCS nonfuel mineral prospecting regulations contain a 20-year period during which resource information generated by private firms will be kept confidential.[14]

A wide range of OCS nonfuel minerals exist, and some already make contributions to national mineral supply. In comparison with hydrocarbon production and reserves, however, the economic importance of the OCS nonfuels is trivial. Considering the early stage of development for the majority of these minerals, even small steps taken to generate information could make important gains toward a fuller understanding of their resource potential. Management concerns, therefore, might usefully be directed at the rate and flow of information.

It is apparent that interest groups and public agencies alike have been involved in positioning themselves for the distribution of benefits that could flow to firms and managers from a working system for the disposal of OCS lands for nonfuel mineral exploration and development. To some degree, the activities of competing interests or agencies have helped to push issues surrounding the management of OCS nonfuels higher up on the nation's policy agenda than they might be if their position was based solely upon a sober analysis of the economic potential of these minerals. This increased public exposure has helped to expand the stock of knowledge about OCS nonfuels and may yet result in a more realistic appraisal of the likely potential and value of these minerals as public assets.

Chapter 8

Management and Research

Old issues never quite die in the management of the Outer Continental Shelf, and new issues constantly emerge. As earlier chapters make clear, the system for managing the OCS lands is a loosely knit cloth of federal and state agencies, local governments, and interest groups with the Minerals Management Service as the major agency with power to implement policy decisions.

This chapter presents recommendations for policy and for research. Prior to such recommendations, however, it is useful to restore the human face to the abstract analyses of preceding chapters. This chapter begins with interviews with two of the top managers who have implemented and influenced many policy decisions made for the Outer Continental Shelf.

Policies are developed and implemented by real people faced with decisions to be made in real time. Two of these people are Bill Bettenberg, former director of the Minerals Management Service, and Carolita Kallaur, deputy associate director of the Minerals Management Service. The following is taken from an interview with these people shortly after the five-year plan of the Minerals Management Service was accepted by Congress, and after litigation had been filed by several states and environmental groups to stop implementation of the plan.[1]

THERE IS NO EXHAUSTION OF ISSUES

William (Bill) Bettenberg is a career employee of the Department of the Interior. He started with the Bureau of Mines in 1964 as an analyst and has since held positions that include director of the Budget Office of the Department of the Interior, acting assistant secretary for Energy and Minerals, and deputy assistant secretary for Policy, Budget, and Administration prior to becoming director of the Minerals Management Service. The following interview took place during his tenure as director in September 1987. In 1988 he stepped down to become associate director of the Minerals Management Service.

S. Farrow (SF): What do you think the Minerals Management Service does particularly well, and a few things that need beefing up?

W. Bettenberg (WB): When we discover something novel, something that is a potential problem, I think we are good in terms of getting on top of it, understanding it, and dealing with it. A couple of examples. Just a couple of years ago, in our environmental studies program, some of the investigators discovered a site in the Gulf of Mexico that was growing tube worms, and all sorts of interesting critters in a place that no one had ever conceived that you might have these critters. They had been discovered, probably within the decade, at certain thermal vents in the bottom of the ocean—the midocean ridges, the Gorda Ridge, and that sort of thing—and here you had the same kinds of critters, which looked pretty rare and unusual, and we put in a program almost immediately for photo documentation to see if there was any more of it. What we since discovered is that it is associated with shallow gas and oil deposits in old, submerged deltas, and deltaic fans and these things are simply feeding off of hydrocarbons that are seeping out. They are all over the place in those fans in the Gulf, and we realized we did not need to continue the photo survey program and we have called it off again.

If somebody says there is a problem, we look at it, but we're not very adept at all about getting the science and the substance heard. You know, perhaps, instead of being 97 percent substance and 3 percent outreach, we ought to be just the exact opposite.

SF: That leads to several questions on information. There have been a number of major decisions recently—reactions to changes in the oil market such as canceling sales, the five-year plan coming out, and suspension of operations. What are the show-stopping pieces of information that are particularly effective in getting an issue heard?

WB: I am not sure what they are. A major part of the problem is not only getting them heard, but getting them heard by people who are interested in hearing. We had some suits filed on the five-year program. Several years ago when we were fairly fresh into the process of evolving the next five-year program, we had a new undersecretary. I briefed her, and went through this entire process. I got to the point where we adopted the five-year program, and she said, "Well, then what happens?" and I said, "Well then we go to court," and she was taken aback. I said, "Well, the Andrus (secretary of the Interior, 1977–1981) program was litigated, the Watt (secretary of the Interior, 1981–1983) program was litigated. We know that this will be litigated. We have to take that into account in the way we go about the process—how we write every document."

SF: This last five-year plan took something like two and a half years to prepare, and for the original plan, Congress allocated nine months. Do you have any suggestions or comments on the process or content that it takes so long?

WB: A couple. This one we had designed to take two years, and we ran the train on schedule, up until the time that Hodel (secretary of the Interior, 1985–1989) came. As the new secretary, at first he wanted to step back and take a look at it—raised the issue of subarea deferrals—and that brought a whole new chapter and dimension to the process. It is one thing to examine and compare and contrast 24 regions. It is another thing to within that, have a hundred proposed subarea deferrals all being examined at the same time.

SF: As an analyst, I have a lot of sympathy with that.

WB: You can imagine the orders of magnitude increasing in complexity. We were ready to go in December, but he (Hodel) wanted to wait until Congress was back, out of concern that if we put it out for the congressional review, and it was a period when they were not in session, some might claim foul. And so there was some wheel-spinning time. And then it was complicated by the legislation on California that built, not only another magnitude of complexity, but we were doing almost a tract-by-tract kind of analysis, with stipulations, in California. All of this was in the context of a separate review process with Congress. This built another five or six months into it, at least.

SF: Do you think this is a one-way trend of increasing complexity?

WB: It probably is. I suspect we need some reform of the process. There was some complexity added to it because of the level of the sophistication of our thinking about it, and I suspect in some cases we got to levels of sophistication that were far overpowering the problem at hand. An example of that is, there was a shift from resources estimated to be in a planning area, not only that were economic, but were leased, with the notion that not all the economic resources would be leased, because companies would not know where all of them were and will overlook some. That was a level of sophistication that added complexity, required a fair amount more work, made the whole process a little less explainable to the typical person, and the differences were not sufficiently great that you would ever sway a decision on it.

SB: Therefore, at this early stage. . . .

WB: You really should be making a gross decision. The requirement for an environmental impact statement (EIS), at that stage of the process, seems to me to be a redundant thing, in that you are talking in extremely hypothetical terms, and you are going to have area by area EISs for the sales, where you can get down and talk more concretely. And in any frontier area, if you have discoveries, you are going to have an EIS, or one or more EISs on developments, where you are talking about real things like a platform, or a group of platforms in a place.

SF: Are there objectives of players who are important in this game that I or others might not be thinking of?

WB: Well, there are clearly some people who are just simply opposed to all offshore leasing and if you can satisfy them on a particular issue, they will simply shift issues.

SF: And there is no exhaustion of issues?

WB: There is no exhaustion of issues, and so that is certainly one group that is out there.

SF: You've had high positions in Interior, across several different administrations, and I am wondering what you see as the consistent part of the program across administrations because the bulge in leasing starting in '83 shows up dramatically under this administration.

WB: Well, the notion of areawide leasing (a large increase in leases offered) was really being developed by program professionals in '79 and '80, and there was a push for larger and larger sales. From an economic standpoint, we clearly had a large inventory of valuable prospects that were being held off the market. The areawide approach was a very rational response, both to that aspect of it, as well as all the environmental information that was accumulated.

SF: And what about as a response to revenue pressures as well?

WB: I have never seen the revenue pressure. Since I have been director I would say that we had not made a single decision, with regard to the timing and size of a sale, with an eye to "How much revenue is this going to get us?" and "Do we need to get it now?"

SF: What about issues broader than a single sale? I would think that OMB (Office of Management and Budget), for their projections. . . .

WB: OMB was really enthusiastic about more money, but that did not drive it. What drove it was the notion that there is a lot of good potential out there, that it will take you forever to get to if you keep holding very small sales.

SF: In the reverse situation, for instance, the projections for Fiscal Year '88 that predict revenues dropping down to 2.8 billion, is there a similar lack of pressure?

WB: Yes.

SF: I find that interesting.

WB: Yes, those revenues are strictly transfer payments.

SF: But they are spendable by the government. Economically, they are transfer payments, but somebody gets to spend them here.

WB: I do not see any pressure at all. The only pressure I have seen is in regards to fair market value. The fair market value argument has primarily, perhaps exclusively, but primarily been raised by those who were opposed to leasing in the first place, and so it is just another one of those issues to raise, because they will grab for whatever issue is available.

SF: And that is in OCSLA (Outer Continental Shelf Lands Act). . . .

WB: Yes. You know, it is a politician discovering something too good to be true. You can go out and give a speech in terms of fiscal integrity while opposing leasing at the same time. All the economic evidence we have is that we do not have a problem in that area. As a matter of fact, we have some rather stellar cases where we have rejected tracts and lost money as a result.

SF: Of course, rejecting particular tracts may have an aggregate effect in the sense that even if you lose money by occasionally rejecting tracts, you are affecting the whole bidding process out there and that issue is probably larger.

WB: Right. We will never really know.

SF: Maybe I am being a little picky here, but I want to ask a last point about what is typical across administrations. In the Fiscal Year '86 report, there was a vacancy for the chief of Monitoring and Penalties. Now, clearly vacancies occur periodically in any organization, but one might ask, as with the big bulge in leasing since '83, "Is that independent of administration?"

WB: Yes. More attention has been put into the effort of reviewing the inspection program in this administration, but I would not describe it as an administration initiative. It is more of a personal initiative. When I was acting director for several months in 1982, I had to come up with an organizational structure for the Minerals Management Service, and so I am the one who basically built the organization chart. In the brief spell that I was here—when I was reviewing their program from a budgetary standpoint and policy standpoint—it looked to me like not enough attention was given to that area. For instance, in the onshore area they were conducting the first-ever class for inspectors. Before that, the way you became an inspector was that you were hired because you had some technical experience in this area, and you went around with an inspector for awhile, and after awhile somebody said, "Yes, this guy knows enough to do it on his own," and that was the entire process.

SF: I've been offshore with inspection teams, and they seem to work hard at it.

WB: Yes. There was no organization specifically responsible for that. I said, "In the weeks that I am going to be here, there is no way that I am going to get a program launched for thinking carefully about inspections, what to inspect and what you do not inspect, how you sample, how you train your people for the inspection, and so forth." So, I just established two boxes in the organizational chart, an office of offshore inspection and an office of onshore inspection, within those hierarchies. . . .

SF: And let them fill them.

WB: And let them fill them. I figured that the most I can do is to make somebody responsible for it. If they put somebody in there, they will feel compelled to do something, and you know, it worked.

SF: Almost like the environmental groups out there—I feel that I cannot exhaust all the issues here in a brief talk with you. Are there particular things that I have not brought up, that you think are of particular interest?

WB: There are just lots and lots of interesting issues. From a public policy perspective, probably the most interesting is the small role of real information on substance.

SF: That is, small role in decision making?

WB: In terms of the debate. It is a journalistic type of debate, and a very political debate, and is almost devoid of substantive content. Years ago, when I was doing graduate work in political science there was a study I reviewed on American voting behavior. It was based on a lot of statistical work being done at the University of Michigan on attitudes and interviews. They defined an educated voter as a person who not only could identify a candidate and an issue, but could correctly tie the issue to the candidate.

SF: I might have to plead guilty on that.

WB: Under their definition, they concluded that something like 8–10 percent of the voters were educated, in that they could tie an issue to a candidate. A friend of mine on the Hill (associated with Congress) told me once when I was chatting with him about an issue, "Oh, you are right on. This is the key issue," he said. This guy is a folksy guy and a very bright analyst. He said, "Bill, what you need is a good slogan."

I said, "What do you mean?" He said, "Bill, a good slogan will beat good analysis every day of the week on the Hill."

SF: Probably not just on the Hill.

WB: Definitely not just on the Hill. To give you an example of that—something that is a public issue at this point—is the set of legal issues that are tied in with coastal zone management and the consistency doctrine. In late May we waded into the specific debate with regard to the California Coastal Commission operating outside of the law.

SF: That created quite a stir.

WB: It has. I had started work on that in the fall of 1986. The staff was coming to me and saying, "The Coastal Commission has done the following, and according to the Coastal Zone Management Act, or the Coastal Commission's own plan, or the regulations of Commerce, or the regulations of the California Coastal Commission, they are not supposed to be able to do this," and they wanted action, that the Commission's actions were unconscionable. So I said, "Okay, you've got to build me the case. I want the allegations and all the supporting evidence, allegation by allegation." So we sent that over to Commerce. They (Commerce) were apparently persuaded by our evidence or other evidence they had—I have never had any conversation with them about the substance of it. Commerce started the process that allows the state to comment, but it is a process, presumably, that would lead the state to reform, or face decertification. A whole spate of editorials said we were doing this because we are the lackeys of big oil, and if we have our way with this it will destroy the Coastal Zone Management program. All of our coasts will be at risk sort of language.

SF: Yes. . . .

WB: Not a single thing I've seen said, "Well, let's see, their first allegation was, and their evidence was, I wonder what the facts are here," or "Those facts are wrong for the following reasons." None of that. I think the *Los Angeles Times* wrote an editorial like that in June, too. They have had some reporters on the trail of the Coastal Commission and had a headline, "The Coastal Commission, An Ideal Gone Astray. Now fifteen years old, panel has become entangled in politics, money, back-room maneuvering."[2] It addresses at least a couple of the substantive issues. It never mentions us. It is as if we never made a ripple in terms of substance. Some of the specific allegations we had were that we did not think that a commission could function when the appointees to it were "at the pleasure of." Since we did that they had a vote where a fellow was tapped on the shoulder, handed a letter thanking him for his services, and removed. He had no warning whatsoever, and it was minutes before the vote, and he was replaced with somebody else who voted against the project. That was in July.

WB: Finally, I am reminded of Henry Kaiser who, as I recall, was once quoted as saying that he knew it was going to be difficult to enter the automobile market after World War II, but he never dreamed that he could drop a billion dollars and never make a ripple. It just takes your breath away how you can logically build a case with allegation particulars and instead the issue becomes your audacity in challenging this cherished institution which is saving the shores, and have nothing to do with any of the issues that you raised.

EQUATIONS WITH AN EMOTIONAL FACTOR

Carolita Kallaur has worked for the Department of the Interior since 1968. During a 10-year stint with the Bureau of Land Management, she rose to the chief of the branch of economic analysis in the Division of Policy Analysis and Development. From there she eventually held positions as assistant director of the Office of OCS Program Co-ordination in the Department of the Interior and chief of the Offshore Leasing Management Division of the Minerals Management Service prior to becoming deputy associate director for Offshore Leasing.

S. Farrow (SF): What do you think are the few key things that an interested person ought to know about the OCS leasing program?

C. Kallaur (CK): Initially, the production aspects . . . the fact that 25 percent of our domestic natural gas production and 12 percent of our oil come from the OCS and the fact that it has a very good environmental record. From experience that we have had over the last 10 to 12 years, we know that coastal residents who are not accustomed to having oil and gas development in their backyard have a number of concerns. We have tried to set up a system so that we understand their concerns, work with them, and share information, but we realize that, at times, the balance that you might strike on the national level is different than the balance you might strike on the regional or local level.

SF: But that is hard for people to accept in those regions.

CK: Yes. I can try to put myself in the position of somebody who is living in New England, or living in Florida, or living in California. But, I think it is important that we also take a larger view. We all use petroleum products, and this source of oil and gas has a very good environmental record. It is important to not just think back to Santa Barbara, and think, my God, we are going to have oily beaches and things, but look in terms of improvements we have made in the way the operations are conducted.

SF: You've been a part of this program for a long time, including its transition when it was distributed out among several different agencies. What do you think the current Minerals Management Service does particularly well now?

CK: I think where we have made some improvements is in the area of analysis. I think it is much easier to work together to try to solve the problems if everyone is part of one agency rather than if you are working with two separate agencies; and I think the quality of our decision documents is high. There are areas where we might make some improvements. Right now, we are focusing our attention on EISs, which are documents that the lay person, and even people here, do not read as a matter of course. . . .

SF: Only if I have to.

CK: We are trying to see whether or not we can improve the readability of these documents, not only so that they are easy to read, but also so they provide the right information for decision makers.

SF: That links into the Environmental Assessment Program—do you see that evolving in any particular way?

CK: Yes, very much so. If you look at the Environmental Assessment Program in a broad way—including environmental studies—we have a major policy change under-way right now. We are trying to redirect the money that we are spending in the studies program more to the postlease side of the house, as opposed to the prelease side of the house. We have spent about $435 million on studies. Much of that has been in learning about the basic environment, and information that would be useful for prelease deci-sions. You do not always have to go and regather that information—once you have it, you have it. In some instances, you may have to go back and take some more samples, but we can draw upon that bank of information. We are now trying to learn more about what's really happened to the environment in the areas where we have had oil and gas operations underway for a number of years. And that would mean spending more money in the Pacific and spending more money in the Gulf of Mexico—the areas where pro-duction has occurred.

SF: As an analyst myself I would think that is a key question, and perhaps a risky one for you and the environmental groups in the sense that, in the five-year plan the social costs are relatively small compared to the benefits of production. Those estimates were just based on discrete events, on spill analysis, and now you are opening up the pos-sibility of persistent effects. Do you see that as a risky approach?

CK: Well, if there is something wrong out there, I think we need to know. We have a scientific committee that has been very active over the last couple of years, helping us reformulate our studies program. Among scientific communities they think there's quite a bit of knowledge on short-term effects. Most scientists agree that the short-term effects from oil and gas operations are pretty negligible. If there is a data gap, it's more in the area of cumulative effects so it's appropriate, both from a scientific and a decision point of view, to focus our attention there.

SF: Another change seems to be more emphasis on task forces, whether with specific states or other agencies, and perhaps more hard mineral oriented than oil and gas. I am wondering to what extent you see those task forces as mixing science and politics when you get different groups together?

CK: Well, science and politics are mixed throughout this program, and throughout the history of this program. It is a very political program, for everything you do has a political flavor, even though it can be scientific or analytic.

Certainly, the program is never static. Because we do operate in a political environ-ment, we have a lot of external pressures on us, and we are smart enough to realize that we still have some problems. We are planning to have a successful program in some of the frontier areas, such as the North Atlantic, Florida, and Alaska, but we recognize that we may have to do things differently. We have looked at the hard mineral program to see if any of that experience is relevant. The secretary recently made a proposal to Governor Martinez of Florida on the South Florida area, to see whether or not he wanted to have a task force with Interior to look at some of the environmental issues that were of concern to the state.

SF: And the same might happen in the North Atlantic?

CK: The North Atlantic is an area where we have had leasing moratoria in effect for a number of years on a portion of the planning area. There was some sense in Congress that we should look to see if there were ways to resolve the concerns in this area. We had a federal register notice this summer that asked for different ideas. Right now we are looking at different possibilities from trying to do what we call an IRM (Institute for Resource Management) program—where you have an outside facilitator bring people together and most likely exclude the governmental entities that are warring over these issues.

SF: Of course, there are different views on the success of IRMs efforts up in Alaska (where interested parties convened and provided a recommendation to Secretary Hodel).

CK: Yes, I think there has been a problem because the secretary did not accept the proposal in its entirety. There has been some sense that it was a failure, and that somehow we did not recognize the efforts of the group. We do not think that is true at all; certainly, a lot of the information that was submitted by the group helped us focus on where we are going to have our sales in the future because we have cut back the planning area considerably. There is also a problem, and I think IRM will recognize this, in that they were doing this off on their own and they were not that well versed in our planning processes, which, for both legal and policy reasons, is pretty well structured. The result was their proposal wasn't submitted at exactly the right time. Also, all they gave us was a map with little or no supporting information. Under both Section 18 and Section 19 of the OCS Lands Act, the secretary has to make balancing decisions. We had some conflicting views from other parties at the same time we got the IRM proposals, so we couldn't say, "Well this map from IRM is it—ignore everyone else."

SF: Especially if the map is a consensus political proposal among that group of people who were there, with varying degrees of documentation. I can imagine it might be difficult to provide documentation to support balancing in such a situation.

CK: Well, in some cases you cannot. It is very hard to go and write out how you came to a certain decision that was really a judgment call, but at least we know that they have information that shows the relative priorities on a different block, and that would help. But we are continuing to work with IRM. And I think, too, we both have learned a lot from talking with each other, and they realize that if we were to do something like that for the North Atlantic, we would try to structure it so that any advice that would come in at any appropriate time—we could use it in our decision process and [it] would also be reflected in the EIS, so that we would be able to meet our balancing requirements. We are still looking at that and looking at some other alternatives in the North Atlantic. We also have had some discussion about how we might deal with Washington and Oregon, because that is an area where we have not had any lease sales since the sixties. So, we are departing from our standard way of doing business because we recognize that it is not perfect yet and there may be a better way of working out these differences.

SF: You are a high-placed person in the agency here, also a high-placed woman in the agency. Do you have any particular insights for women trying to get ahead in management positions or policy positions, in a male-dominated industry or technical agency, such as the Minerals Management Service?

CK: I do not have a really technical background . . . I mean, I studied economics.

SF: Economics is not technical?

CK: Well, it is, but I ended up, just because of the way my career evolved becoming more of a generalist. Certainly, I deal with a lot of technical issues, but right now there are people who work for me who are really the experts in that area. I think the OCS program, compared to other government programs, has a higher number of professional women at higher levels. It may be that it is easier for a woman to do well at a policy level than in some of these traditionally male dominated technical fields. I think it is even getting to the point where poeple out in the field accept the fact that there are women who are going to go out and do the field work. And, I think it's a question of just competence—working hard—regardless if you are a man or a woman. I think that people will accept you if you do a good job, and that times have changed enough so that I don't think it is a problem being a woman, even if I were working in an oil company.

SF: Are there any particular difficulties or successes in managing the program that you'd like to bring up?

CK: One thing that is really important is the question of risk assessment. We have a difficult time in being able to communicate with the public at large about this question of risk assessment—about how everything we do in our lives involves some risk, and that people need to look honestly at this question. Concern about oil and gas leaking off their shores is legitimate, but they should be aware that the risk of this happening is really low.

That gets into another issue: "How do you make this technical program understandable to the intelligent lay person, so that they can make an informed decision about whether or not they should write to their Congressman saying that this is okay or this is not okay?" It is a very difficult issue. I don't think you can compare the program to nuclear power development, or toxic waste, but it is certainly in that area of, "When you have a technical program, how do you get it down into terms so someone can really grasp what you are talking about?"

SF: What little of the psychological literature I have read is that very few people make decisions in risky environments as analysts model them to make those decisions. The literature demonstrates I think that people heavily weight rare, but large, events, and significantly understate probabilities of exposing themselves to some other risks, as when driving drunk. There is a gap between how people perceive events and what scientific evidence might be available. Is this the issue? ·

CK: It is. Because, certainly the work we did in the five-year program is a social cost, and a lot of that risk is definite. That is not something that is understandable or acceptable. There are some other equations that you have to use that have an emotional factor.

These interviews and the analyses of previous chapters emphasize that much remains to be done to improve policy-oriented research for the management of the Outer Continental Shelf. There are many exciting areas for further study—areas that are heavily surveyed in the text as well as topics given slight emphasis in a survey of this kind. The following section suggests what some of these areas for research may be.

RECOMMENDATIONS FOR POLICY AND RESEARCH

Policy

Chapters 3 through 7 contain strong policy recommendations for the management of the Outer Continental Shelf. At the same time, new issues are constantly evolving that require further research. As catastrophic spills such as off Santa Barbara and the *Exxon Valdez* indicate, management must be responsive to both existing and poorly foreseen events and their consequences.

Chapter 3 on resources and exploration contains two recommendations. The first is well known but seldom implemented, that information about the cost distribution of resources provides policy-relevant information as the market situation inevitably changes. A subject studied in the text more extensively is the validity of the government's proposition that there are large external benefits of information from exploration in frontier areas. An investigation of this topic for one area reveals that there are only small differences in expected value when the current process is compared with complete government control of exploration. Until further evidence is available, there is little reason to offer incentives to firms to explore frontier areas based on the expected value of the information externality.

Environmental issues presented in Chapter 4 are more likely targets for research than for policy recommendations at this time, particularly in light of the issues raised and the large effect on recent observations of the *Exxon Valdez* spill. For instance, contemporary accounts indicate that the clean up cost per barrel from that spill may be approximately 10 times larger than previous costs. Research recommendations about the environment are discussed in the following section.

In regard to fair market value, the topic of Chapter 5, one recommendation is that an ongoing validation program be implemented to assess the bias, if any, in economic estimates of the Minerals Management Service and to monitor, perhaps by sampling, the rates of return as a separate check on fair market value. This recommendation does not mean that there is a substantial problem in the area of fair market value. On the contrary, research indicates that, based on current knowledge, there is little problem in this area in aggregate. Nonetheless, the contentiousness and longevity of this issue warrants a monitoring program. A second recommendation is to investigate ways in which state and interest groups could participate directly in the auction process. The existing auction mechanism is well suited to an adaptation such as lease delay rights. Questions about who should pay for the right to explore can be addressed explicitly by creating a rate of exchange between state and interest group bids for lease delay rights and development bids for exploration rights.

The pace of leasing, presented in Chapter 6, remains a contentious issue due to the lack of consensus on both the theory of the pace of leasing in a world market as well as the difficulty of implementing the statutory balancing requirements of OCSLA. One recommendation here is that the decision-making process be made more transparent. The poor relationship between the voluminous data generated and the resulting lease schedule is a strong indication of implicit decision criteria. Specifically, an explicit decision-making process should be developed in more depth that corresponds to the large expenditure of time and effort in generating input data. Second, continual monitoring of the program through integrated financial statements would provide a useful framework to study the trade-off between current income and long-term assets.

In the area of seabed minerals, the developing concerns of this activity can be addressed by allowing flexibility in the exploration process. However, the arguments for strategic materials are a weak base for nonfuel OCS policy.

More generally, managerial improvement cannot be measured without comparing expectations and realizations. Each of the quantitative policy analyses contains the seed of a test between projections and reality. The Minerals Management Service should develop an ongoing, probably computerized, process that generates forecasts for key measures relating to the balancing criteria specified by Congress. The forecasts can then be compared with what actually occurs. Acknowledging that any forecast is virtually certain to be wrong, the process of developing forecasts sensitizes one to the sources of error so that the learning process is incorporated into the management process. A process of this sort would be a useful start toward a set of performance indicators for the Minerals Management Service.

Research

Recommendations for research are subject to one's self-interest. As Nobel Laureate George Stigler warned researchers, "Promising ideas are all that even a rich scholar possesses. . . . revealed preference is the only reliable guide to what a scholar believes to be fruitful research problems."[3]

There are four questions that get to the heart of policy issues on the OCS: (1) What are the effects of persistent, low levels of pollution? (2) What measures of environmental costs from spills should be more carefully studied? (3) What are the differences between measures of risk and the perceptions of risk? (4) Should managers be more cautious about the pace of leasing due to depleting offshore resources?

Recommendations for research about the environment support the recent reordering of priorities in the environmental studies program of the Minerals Management Service. In particular, the effects of spills may be more important in galvanizing public opinion than in their long-term effect (a topic discussed below as risk perception.) Although not ignoring research on undesired experiments such as the spill from the *Exxon Valdez*, scientific research on long-term effects from persistent levels of pollution associated with OCS activities seems indicated as a missing component of current analysis.

In contrast to persistent, low levels of pollution, highly visible spills such as the *Exxon Valdez*, the *Argo Merchant*, and the *Amoco Cadiz* generate intensive study in part due to the litigation that follows. For instance, within two months of the *Exxon Valdez* spill, the Exxon Corporation was committed to spending $15 million for research on damage assessment with preliminary plans to spend a total of $35 million, primarily for the assessment of environmental costs.

Efforts to improve the estimates of the environmental costs from OCS operations by updating information used and by refining and extending the analysis are needed. Improved analysis of oil spill costs could begin by incorporating and extending the results of newly available non-market valuation studies of sport fishing, beach use and other marine resource activities. Important examples include recent work carried out under various Environmental Protection Agency Cooperative agreements, focusing on sport-fishing and other marine topics, and ongoing research by the National Oceanic and Atmospheric Administration that uses a consistent framework to value beach use at more than 40 sites through the coastal United States.[4]

Efforts to improve the NRDAM/CME are also warranted and supported by a court decision in July, 1989 that remanded the damage assessment regulations back to the Department of Interior.[5] For example, the first version of the NRDAM/CME simulated the effects of a spill within a single environment type such as a mud bottom, subtidal environment in a given area for a given season. The model allowed for many environment-season types; but once set, the model assumed that all environmental conditions applied throughout the entire spill. The use of a geographic information system approach and a generally more sophisticated framework could allow environmental conditions to change over space and time throughout the course of the spill.

A particularly challenging task in environmental cost estimation and in revising the NRDAM/CME concerns the possible adoption of credible estimates of so called non-use values such as option and existence value. The measurement of non-use values is a contentious area in economics, and hence a subject needing further research, in part because of the inability to rely on market transactions in order to estimate and validate non-use values.[6] The work described in Chapter 4 includes only lost use value; to the extent non-use values are important in some cases, including credible estimates of these possible costs would provide a more complete picture of the environmental costs of OCS oil and gas leasing.

Another potential but little researched cost of the OCS program is the effect of off-shore drilling on onshore property values.[7] Much of the opposition to OCS development off California and Massachusetts is possibly the result of extensive onshore development that has capitalized ocean views and amenities without oil rigs into the value of the land. The opposition to leasing may be a concern for a loss in wealth about which a substantial body of theoretical literature and empirical work already exists. New research could investigate the value of properties with and without oil rigs while controlling for other factors (such as distance from the ocean and size of the house) through regression analysis. Although research in the Minerals Management Service has begun to look at this effect on attendance at public beaches, the work was preliminary and does not exhaust the topic.[8]

A complementary question to whether analysis matters as presented in Chapter 6, is whether the appropriate analyses have been conducted. While this issue appeared to have been settled by litigation over the five-year plan and in many individual sales through litigation over environmental impact statements, it is re-emerging as a topic of debate in regard to individual lease sales. A recent report by the National Research Council, prepared as an advisory document for a multi-agency review committee investigating leasing off Florida and California, concluded that information was inadequate for informed decision making in those areas.[9] The report stated that socioeconomic information in particular has been too narrowly construed and too little research has been conducted. In at least one leasing area, the report also concluded that ecologic and physical oceanographic information was inadequate. It remains to be seen whether the National Research Council's interpretation of the adequacy of information, an interpretation complicated by vague statutory guidelines and poorly defined decision criteria, will result in an altered emphasis or expanded funding for OCS research. In any event, the report will lead to further debate about the content and use of information in OCS decisions.

A different category of research relates to the perception held by the public about the risk related to OCS operations. The role that risk perception plays in focusing support or opposition to the program is largely unexplored. Recent research in other areas indicates that many subjective factors in the perception of risk seem to correspond to concern about environmental issues and oil spills on the Outer Continental Shelf.[10]

The literature about risk perception indicates that in many instances people assign too high a probability to rare but highly visible events. Many new topics for research relate to this issue, such as the extent to which oil spills are analogous to other risky events and the characteristics of oil spills that are particularly important to the public. In other areas, psychologists have identified characteristics such as "dread" and "choice" as elements that affect risk perceptions. A careful study of risk perceptions may also provide further information about the social costs of the OCS program.

Last among the proposed research areas is the characterization and the quantification of a depletion effect from offshore exploration and production. The economics literature is less concerned with total depletion of a resource than with the way in which markets include, or fail to include, information about depletion. Although there are theoretical discussions of a depletion effect, there have been few empirical applications.[11]

In general, economists believe that economic depletion often causes exploration and extraction to come from more expensive sources, an indirect cost which a market might ignore. Definition of this depletion effect in both frontier and mature areas of the OCS, as well as its empirical estimation, could provide a basis for changing the bid adequacy procedures of the Minerals Management Service if, in fact, an additional cost item should be added to the expected value of each tract.

Like exploration for new resource deposits, major research findings and new policy suggestions can come from new ways of looking at a problem. What exactly will new research and policy for the Outer Continental Shelf lands contribute to the welfare of the country? No one knows precisely but the contribution will be large given the richness of the ocean environment, the hunger for energy within the economy and the location of over 40 percent of the population within 50 miles of ocean coasts. Managing the OCS lands will continue to require steering through an ocean of controversy in the face of changing conditions.

Appendix:

Management Information and
Information Sources

The Minerals Management service generates a wealth of information, much of which, though not all, is available to the public. This appendix identifies the key annual publications and the location of information offices as well as a brief discussion of data bases in the Minerals Management Service.

Data Bases

As of 1985 the Minerals Management Service had developed and maintained 15 major and approximately 30 minor computerized data and analysis systems spread throughout the regions and headquarters. These systems and the personnel associated with their use are estimated to account for 10 percent of the budget of the Minerals Management Service.[1]

Information on preleasing and the results of the auction and bid adequacy procedures are maintained in a data base called Post Sale Analysis System (PSAS or POSTSALE). This data base includes information on each tract from the auction such as the value of any bids received, the number of bidders, the estimated resources, and the value as estimated by the Minerals Management Service. Some of these data are publicly available, such as the bids and the Minerals Management Service evaluation for leased tracts, and may be generated by subsidiary computer programs such as those to estimate resources and value. The POSTSALE data base is used to prepare extensive internal reports immediately following a lease sale.

A second data base is maintained for postlease exploration and production records. Called Outer Continental Shelf Information System (OCSIS), it contains information on each tract such as the time and depth of each well drilled, the number of production platforms, and the amount of production of oil and gas. Information from POSTSALE, OCSIS, and some additional sources of information are used to form a third data base called Offshore Lease Data System (OLDS), which contains approximately 150 pieces of information on a tract-specific basis for leases evaluated or bid on and their subsequent exploration and production history. Again, there is a mixture of available and proprietary data in OLDS. Other major systems are presented in Figure A-1.

Not listed, however, is a proposed data base and graphics system for management that may link several systems to create the Offshore Policy and Management Information System (OPMIS). Important data bases and computer models exist in other agencies as well. An example is the Type A damage assessment model for quantifying damages from the spilling of hazardous wastes and oil into the ocean.[2]

Computerized Data and Analysis Systems

Outer Continental Shelf Information System (OCSIS)
Offshore Inspection System (OIS)
Offshore Lease Data System (OLDS)
Post Sale Analysis System (PSAS)
Two-Stage Leasing Model (TSL80—used to value prospects in the Five Year Leasing Plan)
Probabilistic Resource Estimates-Offshore (PRESTO—used to estimate unleased resources in the five-year plan and discussed in Chapter 4)
Monte Carlo (MONTCAR—used to conduct the bid adequacy review)
Marine Survey System (MARS)
Oil Spill Risk Analysis (OSRA)
Offshore Coastal Dispersion (OCD)
Map Overlay and Statistical System (MOSS)
Automated Cartographic System (ACS)
Geophysical Mapping system (GMS)
Plume Airshed Reactive Interactive System (PARIS)
Well Status and Production System (WSPS)

Major Minerals Management Service Publications

Minerals Management Service, Annual Report (Fiscal Year), "Oil and Gas Leasing/ Production Program." Submitted annually to Congress. Source: OCSIP.
———. "Federal Offshore Statistics: 19—." Annual statistical report. Source: OCSIP.
———. "OCS National Compendium: OCS Oil and Gas Information through 1988." Detailed description of each planning area as well as some data. Periodically revised. Source: OCSIP.
———. "Offshore Scientific and Technical Publications." Issued several times a year; a catalogue of publications. Source: Technical Publications Unit, Herndon.
———. "Regional Summary Report/Index." One for each major region, revised periodically to present descriptions of activities and recent data.
———. "OCS Directory." A listing of the people, the agencies and the committees involved in the oil and gas program. Source: OCSIP.
U.S. Government Printing Office, "Code of Federal Regulations: Title 30, parts 200–299." The current regulations of the Minerals Management Service.

Addresses

Alaska: Library, Alaska OCS Region, Minerals Management Service, 949 East 36 Ave., Room 110, Anchorage, AK 99508-4302.

Atlantic: Public Information Officer, Atlantic OCS Region, Minerals Management Service, 381 Elden St., Herndon, VA 22070-4871.

Gulf of Mexico: Public Information Section, Gulf of Mexico OCS Region, Minerals Management Service, 1201 Elmwood Park Blvd., New Orleans, LA 70123.

Pacific: Public Affairs, Pacific OCS Region, Minerals Management Service, 1340 West 6th St., Los Angeles, CA 90017.

OCS Information Program (OCSIP): OCS Information Program, Office of Offshore Information and Publications, Minerals Management Service, MS 642, 381 Elden St., Herndon, VA 22070-4871.

Technical Publications: Technical Publications Unit, Office of Offshore Information and Publications, Minerals Management Service, MS 642, 381 Elden St., Herndon, VA 22070-4871.

National Oceanographic and Atmospheric Administration, Coastal Zone Management Program, U.S. Department of Commerce, Washington, D.C., 20852.

U.S. Department of the Interior, Office of Public Affairs, 18th and C Sts. N.W., Washington, D.C. 20240.

U.S. Senate Committee on Mineral Resources Development and Production, SD-362 Dirksen Senate Office Bldg., Washington, D.C. 20510. Responsible for oversight of federal mineral leasing and oil and gas production.

U.S. Senate Committee on Energy and Natural Resources, SD-364 Dirksen Senate Office Building, Washington, D.C. 20510.

U.S. Senate Committee on Environmental Protection (a subcommittee of the Environment and Public Works), SD-458 Dirksen Senate Office Building, Washington, D.C. 20510. Responsible for oversight of environmental aspects of OCS lands.

U.S. House of Representatives Committee on the Panama Canal and the Outer Continental Shelf (a subcommittee of Merchant Marine and Fisheries). H2-579 House Office Bldg. Annex II, Washington, D.C. 20515. This committee has primary oversight jurisdiction over the Outer Continental Shelf Lands Act and management of the OCS by the Department of the Interior.

Variable Definitions: Areawide Regression

AAA	Real annual AAA corporate bond rate.
AREAWIDE	1 if the date exceeds May, 1983; 0 otherwise.
AVGPC	Average annual real price change used in the Minerals Management Service evaluations for each sale.
CALYR8X	Dummy variable for the calendar year of sale (198X = 1; 0 otherwise).
PRICE	Real refiner acquisition cost of domestic crude oil.
RMROV	Real MROV (mean range of value) in constant 1984 dollars.
RHIBID	Real high bid in constant 1984 dollars.

Notes

Chapter 1

1. M. Boskin, M. Robinson, T. O'Reilly and P. Kumar, "New Estimates of the Value of Federal Mineral Rights and Land," *American Economic Review*, December 1985.

2. D. Rosenthal, M. Rose, and L. Slaski, "The Economic Value of the Oil and Gas Resources on the Outer Continental Shelf," *Marine Resource Economics*, 1989, pp. 171–189.

3. See P. Gates, *History of Public Land Use*, Federal Land Law Commission, 1969, and M. Clawson, *The Federal Lands Revisited* (Baltimore: Johns Hopkins Press, 1980).

4. Several insightful and entertaining histories are: J. Leshy, *The Mining Law: A Study in Perpetual Motion* (Baltimore: Johns Hopkins Press, 1986); C. Jayer and G. Riley, *Private Domain, Public Dominion,* (San Francisco: Sierra Club Books, 1985); S. McDonald, *The Leasing of Federal Lands for Fossil Fuels Production,* (Baltimore: Johns Hopkins Press, 1979).

5. A popular account of this evolution is by Roderick Nash, *Wilderness and the American Mind* (New Haven: Yale University Press, 3rd ed. 1982). A second work, mixing the importance of politics, regulation, science and philosophy is *Beyond the Hundredth Meridian* the biography of John Wesley Powell by Wallace Stegner (Boston: Houghton Mifflin, 1954).

6. A useful reference is L. Dye, *Blowout at Platform A,* (Garden City: Doubleday 1971). The 1989 spill from the *Exxon Valdez* in Alaska will have similar long-term repercussions. Within months of that spill, Congress was debating increases in liability limits for companies who spill oil, the Coast Guard reviewed its traffic management facilities, tougher regulations for personnel and ship handling were discussed, and the entire offshore leasing program came under intensive review.

7. Many introductions to geology are available, e.g., R. Chapman, *Petroleum Geology* (Amsterdam: Elsevier Scientific Publishing, 1973); *Fundamentals of Petroleum*, M. Gerding ed., 3rd ed. (Austin: University of Texas 1986).

8. National Academy of Sciences, *Oil in the Sea: Inputs, Fates and Effects* (Washington, D.C.: National Academy Press, 1985).

9. F. P. Shepard, *Submarine Geology,* 3rd ed. (New York: Harper and Row,), p. 198, 1973.

10. A survey of offshore technologies can be found in R. Baker, *A Primer of Offshore Operations,* 2nd ed. (Austin: University of Texas 1985); also H. Whitehead, *An A–Z of Offshore Oil and Gas,* 2nd ed. (Gulf Publishing, 1983).

11. L. Dye, *Blowout at Platform A* (Garden City: Doubleday, 1971); U.S. Geological Survey Professional Paper 679, "Geology, Petroleum Development, and Seismicity of the Santa Barbara Channel Region, California;" U.S. Geological Survey Professional Paper 676, "Geologic Characteristics of the Dos Cuadras Offshore Oil Field," T. H. McCulloh.

12. Increased oil production from Alaska, whether from state lands, from the OCS, or possibly from the Alaska National Wildlife Refuge is expected to be transported from Alaska to the

mainland through the same route as the cargo in the *Exxon Valdez*. Transportation by tanker ship in this area contrasts with transportation in the lower 48 states where pipelines are the primary carrier of offshore oil and gas to domestic refineries.

13. National Research Council, *Oil in the Sea: Inputs, Fates, and Effects* (Washington, D.C.: National Academy Press, 1985), see especially Chapter 5 on effects of oil in the sea. The appendix contains useful case histories of major or well-studied oil spills.

14. Bibliographic information on environmental studies relating to the OCS can be found in National Oceanic and Atmospheric Administration, *"Outer Continental Shelf Environmental Assessment Program Comprehensive Bibliography,"* MMS 88-0002. Proceedings of regional information transfer meetings sponsored by the Minerals Management Service are also useful sources such as *Pacific OCS Region, 1987, Information Transfer Meeting: Proceedings of a conference,* MMS 87-0032. See also NRC, *Oil in the Sea.*

15. See e.g., W. N. Tiffney, Jr., ed., *Proceedings of the 1985 California Offshore Petroleum Conference* (Los Angeles: American Society for Environmental Education, 1985); National Research Council, *The Adequacy of Environmental Information for OCS Oil and Gas Decisions* (Washington, DC: National Academy Press, 1989).

16. U.S. Geological Survey and the Minerals Management Service, *National Assessment of Undiscovered Conventional Oil and Gas Resources,* Working Paper, 1989, p. 138. Complications occur in frontier areas where total resources in an area must be of sufficient size to develop the transportation infrastructure.

Chapter 2

1. A survey of this problem can be found in G. Downs and P. Larkey, *The Search for Government Efficiency* (Philadelphia: Temple University Press, 1985).

2. The Submerged Lands Act, (43 U.S.C. 1311); The Outer Continental Shelf Lands Act (43 U.S.C. 1331–1356); and the Outer Continental Shelf Lands Act Amendment of 1978 (43 U.S.C. 1811).

3. Minerals Management Service, *OCS Laws Related to Mineral Resource Activities on the OCS;* OCS Report MMS85-0069.

4. These issues also appear in Chapter 5 where the congressionally mandated development of a five-year leasing plan is linked to the pace of leasing and to the judicial challenges to the preparation of recent plans.

5. 05 USC 551–559, 701–706.

6. *Cal. v. Watt* (668 F.2d 1290, 1981).

7. M. Weidenbaum, *Business, Government, and the Public* (Englewood Cliffs, NJ: Prentice Hall, 1986); W. Niskanen, *Bureaucracy and Representative Government* (Chicago: Aldine Press, 1971); D. McFadden, "The Revealed Preferences of a Government Bureaucracy," *The Bell Journal of Economics,* Spring 1975.

8. A classic study in this area is H. Kaufman, *The Forest Ranger: A Study in Administrative Behavior* (Baltimore: Johns Hopkins Press, 1960); an example in the private sector is T. Peters and R. Waterman, *In Search of Excellence* (New York: Harper and Row, 1982).

9. See W. Stegner, *Beyond the Hundredth Meridian* (Boston: Houghton Mifflin, 1954).

10. The current division of labor places onshore leasing in the control of the Bureau of Land Management, offshore leasing in the Minerals Management Service, and the collection of both onshore and offshore royalties in the Minerals Management Service.

11. See e.g., M. Olsen, *The Logic of Collective Action* (Cambridge: Harvard University, 1971), or S. Ross, "Organizing for Marine Policy: Some Views from Organization Theory," in *Making Ocean Policy: The Politics of Government Organization and Management,* F. Hoole, R. Friedheim, and T. Hennessey, ed. (Boulder: Westview, 1981).

12. P.L. 85-212 Section 4.2.

13. 43 U.S.C. 1344.

14. Legal challenges to the five-year plan are discussed in Chapter 6.

15. States and local government have broader powers in the postleasing process.

16. 48 CFR 24296.

17. "Offshore Oil Agreement Collapses," *Washington Post,* September 11, 1985.

18. A policy proposal to allow states and interest groups to participate in auctions is discussed in Chapter 5.

19. This is an issue of long standing that has led to unitization laws that can lead to forced merging of production from adjacent tracts both onshore and within a single jurisdiction. See, e.g., Gary Libecap and Steven Wiggins, "The Influence of Private Contractual Failure on Regulation: The Case of Oil Field Unitization, *Journal of Political Economy,* 93:4, 1985.

20. Section 8(g) of the 1978 OCSLA amendments.

21. R. Marzulla, "Federalism Implications of OCSLA Section 8(g)," *Natural Resources and Environment,* Spring 1986.

22. P.L. 99-272.

23. *California vs. Kleppe,* 605 F.2d 1187, 1199 (9th Cir. 1979) In regard to other actions such as the discharge of drilling muds, EPA permitting processes are required. Environmental impact statements are also required for oil and gas activities that are determined to have a significant effect on the environment.

24. *Interior vs. California,* 104 S. Ct. 656 (1984): a useful summary of these issues is in R. Hildreth "Ocean Resources and Intergovernmental Relations in the 80's: Outer Continental Shelf Hydrocarbons and Minerals," in *Ocean Resources and U.S. Intergovernmental Relations in the 1980's,* M. Silva, ed.; and in T. Eichenberg and J. Archer, "The Federal Consistency Doctrine: Coastal Zone Management and "New Federalism," *Ecology Law Quarterly* 14:9–68, 1987.

25. The North Slope is not a part of the OCS, but this topic has also been considered by the OCS policy committee.

26. For example, W. Mead, A. Moseidjord, D. Muraoka and P. Sorenson, *Offshore Lands: Oil and Gas Leasing and Conservation on the Outer Continental Shelf* (San Francisco: Pacific Institute for Public Policy Research, 1985). Department of the Interior, *Five Year Leasing Program,* Appendix F; T. Page, *Conservation and Economic Efficiency* (Baltimore: Johns Hopkins University Press, 1977).

27. See, e.g., "Comments by the Natural Resources Defense Fund, Proposed Five Year Plan," and the response, in part contained in Appendix F of the final five-year plan. Advanced discussions of discounting can be found in P. Dasgupta and G. Heal, *Exhaustible Resources and Economic Theory,* (Cambridge: Cambridge University Press, 1979); and in L. Lind, *Discounting for Time and Risk in Energy Policy Analysis* (Baltimore: Johns Hopkins University Press).

28. As a thought experiment it might be useful to consider whether your purchases would be substantially the same, after a period of adjustment, if you suddenly became much wealthier or poorer. Would any change in purchases be similar to the purchases of those who are currently wealthy or poor?

29. Economists often argue that if the current income distribution is not acceptable to policymakers, this issue should be addressed explicitly by policies to redistribute income on a broad basis instead of on a piecemeal basis on every policy issue.

30. Discussions of these issues can be found in R. Stroup and J. Baden, *Natural Resources: Bureaucratic Myths and Environmental Management* (San Francisco: Pacific Institute for Public Policy Research, 1983) and in S. Brubaker, ed., *Rethinking the Federal Lands.* The focus of these discussions are the onshore lands though the same general principles apply.

Chapter 3

1. Of course, private firms also spend large sums of money on studying the oil and gas resources worldwide.

2. See, e.g., U.S. Bureau of Mines and U.S. Geological Survey, 1980, "Principles of a Resource/Reserve Classification for Minerals," U.S.G.S. Circular 831; T. Tietenberg, *Environmental and Natural Resource Economics,* 2nd ed. (Glenview, IL: Scott, Foresman, 1987).

3. A. Wildavsky and E. Tenenbaum, *The Politics of Mistrust: Estimating American Oil and Gas Resources* (Beverly Hills, CA: Sage Publications, 1981).

4. Taken from Wildavsky and Tenenbaum, *Politics of Mistrust.*

5. L. Cooke, *Estimates of Undiscovered, Economically Recoverable Oil and Gas Resources for the Outer Continental Shelf as of July 1984,* MMS85-0012; USGS-MMS Working Paper, *National Assessment of Undiscovered Conventional Oil and Gas Resources,* Open-File Report 88-373.

6. G. Dolton et al., (1981) *Estimates of undiscovered recoverable conventional resources of oil and gas in the United States,* U.S. Geological Survey Circular 860.

7. Curlin, James, "The Uncertain Future of Offshore Petroleum Resources," mimeo, Marine Policy Center, The Woods Hole Oceanographic Institution, 1987.

8. M. Uman, W. James, and H. Tomlinson, "Oil and Gas in Offshore Tracts: Estimates Before and After Drilling," *Science,* August 3, 1979, pp. 489–491; J. Davis and J. Harbaugh, "Oil and Gas in Offshore Tracts: Inexactness of Resource Estimates Prior to Drilling," *Science,* August 29, 1980, pp. 1047–1048.

9. Uman et al., "Oil and Gas."

10. Davis and Harbaugh, "Oil and Gas."

11. Recall that whereas the mean of the sum of random variables is the sum of the means, the variance of a sum is equal to the sum of the variance only if the variables are independent.

12. Uman, et al., "Reply." *Science,* August 29, 1980, p. 1048.

13. H. Raiffa, *Decision Analysis,* (Reading: Addison-Wesley, 1968).

14. A. Solow, and J. Broadus, "Loss Functions in Estimating Offshore Oil Resources," *Resources and Energy,* March, 1989.

15. Solow and Broadus, "Loss Functions."

16. *Ibid.*

17. The program used to estimate the minimum economic field size, MONTCAR, is also used to evaluate bids from the auctions, as discussed in Chapter 6.

18. Minerals Management Service, 1987, Appendix F-75. A portion of these data are reproduced in Table 4-1 of this text.

19. In unexplored areas, it is also possible for companies to band together to drill an information well that is intentionally not drilled in a prospective site for oil and gas. These strategraphic information wells are called COST wells.

20. Paul Kobrin, "Finding Oil in the Arctic: Three Case Studies of Successful Exploration," American Petroleum Institute, Critique #021, April 1988.

21. This theory is based on an extension of that discussed in M. Mangel, *Decision and Control in Uncertain Resource Systems* (Orlando: Academic Press, 1985), pp. 93–107. More complex theories that also explain the level of spending on exploration are presented in A. Fisher, "Measures of Natural Resource Scarcity," in *Scarcity and Growth Reconsidered,* by V. K. Smith, ed. (Baltimore: Johns Hopkins Press, 1980).

22. This topic is discussed in many sources. Many economic applications are discussed in A. Chiang, *Fundamental Methods of Mathematical Economics* (New York: McGraw Hill, 1974).

23. Recall that the administrative details of the Homestead Act such as acreage limitations may have been a key contributor to violence, fraud, and overgrazing in the frontier West because the details were not adapted to those arid lands.

24. Federal Register, 51:211, pp. 39810–39812 (October 31, 1986).

25. Further detail on the material in this section can be found in S. Farrow and M. Rose, "Public Information Externalities: Estimates from Offshore Energy Exploration," School of Urban and Public Affairs Working paper 88-12 and revision (Pittsburgh: Carnegie Mellon University, 1988).

26. See, e.g., F. Peterson, "Two Externalities in Petroleum Exploration," in *Studies in Energy Tax Policy*, G. Brannon, ed. (Cambridge: Ballinger, 1975); R. Gilbert, "The Social and Private Value of Exploration Information," in *The Economics of Exploration for Energy Resources*, J. Ramsey, ed. (Greenwich, CT: JAI Press, 1981); C. Mason, "Exploration, Information, and Regulation in an Exhaustible Mineral Industry," *Journal of Environmental Economics and Management* 13:153–166, 1986.

27. A more comprehensive definition of public value is discussed in Chapter 4 in connection with environmental studies.

28. Minerals Management Service, Branch of Economic Studies, *The Revenue Model*.

29. The analysis assumes that the sequence of drilling is in the order of expected net benefits. This is consistent with the basic model of search theory presented in the preceding section and casual empiricism of offshore drilling.

30. J. Lohrenz, and E. Dougherty, "Federal Offshore Oil and Gas Lease bonus bid Rejections: Viewpoint of Bidders and Owners," Society of Petroleum Engineers, 11309, Hydrocarbon Economics and Evaluation Symposium (1983).

Chapter 4

1. See, e.g., R. L. Little and L. A. Robbins, "Effects of Renewable Resource Harvest Disruptions on Socioeconomic and Sociocultural Systems: St. Lawrence Island," Technical Report No. 89, Contract No. AA851-CT1-59, Minerals Management Service, U.S. Department of Interior, Anchorage, Alaska, June 1984; J. G. Jorgensen, "Effects of Renewable Resource Harvest Disruptions on Socioeconomic and Sociocultural Systems: Norton Sound," Technical Report No. 90, Contract No. AA851-CT1-59, Minerals Management Service, U.S. Department of Interior, Anchorage, Alaska, January 1984.

2. The possible effects of drilling fluids on the marine environment are not considered. A report by the National Academy of Sciences, "Drilling Discharges in the Marine Environment," 1983, concludes that drilling fluids discharged during exploratory drilling, in general, do not pose a serious threat to the environment (although localized levels of pollution could be temporarily high, and burial of benthic organisms in the vicinity of drilling operations could occur). Major, long-term monitoring studies in progress on the Southern California OCS and similar studies proposed for other areas should greatly enhance understanding of the long-term effects of OCS operations.

3. An alternative measure of regional costs would be the net costs to residents resulting from OCS development in that area. This is the accounting stance adopted in the Massachusetts Institute of Technology (1983) study of prospective OCS oil and gas development on Georges Bank and is a useful framework from a region's point of view. However, when viewing the distribution of the costs and benefits among regions from a national perspective, the regional measure of costs described in the text is most appropriate.

4. See, e.g., W. J. Mead and P. E. Sorensen, "The Economic Cost of the Santa Barbara Oil Spill," paper presented at the Santa Barbara Oil Symposium, Santa Barbara, December 1970; Massachusetts Institute of Technology, "The Georges Bank Petroleum Study," Cambridge: Sea

Grant Report 73-5, 1973; T. A. Grigalunas, "Offshore Petroleum and New England," Kingston, Marine Technical Report No. 39, 1975; U.S. Congress, Office of Technology Assessment, "Working Paper for Coastal Effects of Offshore Energy Systems," Washington, D.C., 1976; Woods Hole Oceanographic Institution, "Effects on Commercial Fishing of Petroleum Development off the Northeastern U.S.," Woods Hole, 1976; University of Rhode Island, "Fishing and Petroleum Interactions on Georges Bank, Vol. II," Energy Program Technical Report 77-1, Kingston, RI, March 1977; University of Aberdeen, "A Physical and Economic Evaluation of Loss of Access to Fishing Grounds Due to Oil and Gas Installations in the North Sea," University of Aberdeen, March 1978; U.S. Department of Commerce, National Oceanographic and Atmospheric Administration, "Assessing the Social Costs of Oil Spills: The AMOCO CADIZ Case Study," U.S. Department of Commerce, NOAA, 1983; J. J. Opaluch and T. A. Grigalunas, "Controlling Stochastic Pollution Events through Liability Rules: Some Evidence from OCS Leasing," *Rand Journal of Economics,* 15 Spring 1984, pp. 142–151; E. A. Wilman "External Costs of Coastal Beach Pollution, An Hedonic Approach," Washington, D.C., Resources for the Future, Inc., 1984; Centaur Associates, "Bering Sea Commercial Fishing Industry Impact Analysis," Technical Report No. 97, Minerals Management Service, U.S. Department of Interior, Anchorage Alaska, April 1984; Grigalunas et al. (1986); T. A. Grigalunas, J. J. Opaluch, D. French, and M. Reed, "Measuring Damages to Coastal and Marine Natural Resources: Concepts and Data Relevant for CERCLA Type A Damage Assessments," Springfield, VA, National Technical Information Service, March 1987.

5. These are starting prices of oil and they are assumed to increase in real terms at a 1 percent rate.

6. See U.S. Department of the Interior (1987, Appendix F).

7. U.S. Department of the Interior, Minerals Management Service, Final Regional Environmental Impact Statement, Gulf of Mexico, Vol. 1, Springfield, VA., National Technical Information Service, January 1983, p. 70; U.S. Department of Transportation. Offshore Oil Pollution Compensation Fund. Report to Congress. Title III, Outer Continental Shelf Lands Act Amendments of 1978. Washington, DC, Fiscal Year 1984; Department of the Interior, *Federal Offshore Statistics: 1988,* OCS Report MMS 89-0082. The spill from the *Exxon Valdez* was not related to OCS production.

8. See, e.g., Mead and Sorenson, "Economic Cost of the Santa Barbara Oil Spill"; T. Grigalunas, R. C. Anderson, G. M. Brown, Jr., R. Conger, W. J. Meade, and P. E. Sorensen, "Measuring the Cost of Oil Spills: Lessons from the AMOCO CADIZ Incident," *Marine Resource Economics,* 2, No. 3, 239–262.

9. K. J. Lanfear and D. E. Amstutz, "A Reexamination of Occurrence Rates for Accidental Oil Spills on the U.S. Outer Continental Shelf," 1983 Oil Spill Conference.

10. U.S. Department of the Interior, Minerals Management Service, Final Regional Environmental Impact Statement, Gulf of Mexico, Vol. 1, Springfield, VA, National Technical Information Service, January 1983, p. 70.; Rainey, personal communication, 1984.

11. Given the fact that most spills are small, use of the mean value results in a higher estimate of the amount spilled than would use of other common measures of central tendency such as the mode or median. Hence, use of the mean value is consistent with the philosophy of overstating environmental costs.

12. Descriptions of the approach used to model oil spill trajectories can be found in U.S. Department of the Interior, 1979, Appendix D. The number used is the probability that a spill that occurs will strike land within 30 days. The estimate for each OCS area was provided by LaBelle (personal communication, 1985).

13. Centaur Associates, "Bering Sea Commercial Fishing Industry Impact Analysis," Technical Report No. 97, Minerals Management Service, U.S. Dept. of Interior, Anchorage, Alaska, 1984, p. 281.

14. NMFS, 1986; Jackson, personal communication, 1986.

15. E. R. Turner and D. R. Cahoon, "Causes of Wetland Losses in the Coastal Central Gulf of Mexico," Minerals Management Service, Jan., 1989.

16. L. A. McMahon, 1984 *Dodge Guide to Public Works and Heavy Construction Costs,* annual ed. No. 16, p. XII (Princeton, NJ: McGraw-Hill, 1983).

17. T. R. Gupta and J. H. Foster, "Economic Criteria for Freshwater Wetland Policy in Massachusetts." *American Journal of Agricultural Economics* 57, 1(1975), 40–45.

18. For example, a site-specific study of the value of Louisiana coastal wetlands by R. Costanza and S. C. Farber, "The Economic Value of Coastal Wetlands in Louisiana" (Louisiana State University, 1985) estimated that the value of an acre of wetlands ranged from $2,429 to $8,977 using a discount rate of 8 percent and 3 percent, respectively. The value per acre used in the text is $13,847, based on the use of an 8 percent discount rate.

19. Mead, and Sorensen, "Economic Cost of the Santa Barbara Oil spill."

20. A. M. Freeman III. "Assessing Damages to Marine Resources: PCB's in New Bedford Harbor," Paper presented at the meetings of the Association of Environmental and Resource Economists and the American Economics Association, Chicago, December 29, 1987.

21. Of course, if a lower rate of discount is used to assess environmental costs, it follows that a lower rate also should be used to quantify the present value of benefits.

22. For a detailed description of the model and data, see Grigalunas et al.

23. U.S. Department of the Interior, Five Year Leasing Plan, Appendix I.

24. R. Just and D. Zilberman, "Asymmetry of Taxes and Subsidies in Regulating Stochastic Mishap," *Quarterly Journal of Economics,* January 1979, pp. 139–148; Opaluch and Grigalunas, "Controlling Stochastic Pollution."

25. J. Conrad, "Oil Spills: Policies for Prevention, Recovery, and Compensation," *Public Policy,* Spring 1980, pp. 143–170.

26. D. Hughart, "Informational Asymmetry, Bidding Strategies, and the Marketing of Offshore Petroleum Leases," *Journal of Political Economy,* January 1979, pp. 139–148; W. Mead, P. Sorensen, and A. Moseijord, "Competitive Bidding Under Asymmetric Information," Community and Organization Research Institute, University of California, Santa Barbara, 1982.

27. See, e.g., D. Reece "Competitive Bidding for Offshore Petroleum Leases," *Bell Journal of Economics,* Autumn 1978, pp. 369–384; M. Rothkopf, "A Model of Rational Competitive Bidding," *Management Science,* March 1969, pp. 362–373; R. Wilson, "A Bidding Model of Perfect Competition," *Review of Economic Studies,* October 1977, pp. 511–518; Mead "Competition and Performance."

28. Opaluch and Grigalunas, "Controlling Stochastic Pollution."

29. U.S. Department of Commerce, "Assessing the Social Costs of Oil Spills"; Grigalunas et al., 1986.

30. T. A. Grigalunas and J. J. Opaluch, "Assessing Damages for Liability under CERCLA: A New Approach for Providing Incentives for Pollution Avoidance?" *Natural Resources Journal,* vol. 28, no. 3 (Summer), pp. 509–533.

31. 30 CFR 250, January 17, 1989.

32. 40 CFR Part 435, October 21, 1989.

Chapter 5

1. 712 F.2d (1983).

2. The most exhaustive recent study of this issue is by the Linowes Commission for onshore coal leases in *Fair Market Value Policy for Federal Coal Leasing,* February 1984.

3. 712 F.2d 584 (1983) 10.

4. Experiments with alternative leasing systems are discussed in W. Mead, A. Moseidjord, D. Mauraoka, and P. Sorensen, *Offshore Lands: Oil and Gas Leasing and Conservation on the Outer Continental Shelf,* Pacific Institute for Public Policy Research, 1985; U.S. General Accounting Office, *Congress Should Extend Mandate to Experiment with Alternative Bidding Systems in Leasing Offshore Lands,* GAO/RCED-83-139, May 1983.

5. Minerals Management Service, *Annual Report of Offshore Oil and Gas Production,* 1985; K. Hendricks and R. Porter, "An Empirical Study of an Auction with Asymmetric Information," *American Economic Review,* December 1988.

6. Five Year Plan-Appendix K, page K-11. There are also provisions for rejecting anomalously low bids.

7. Interior Board of Land Appeals, Decision, 82-706 and Exhibit "B," "An Analysis of Competitive Bids and Current Expected Values for National Petroleum Reserve-Alaska Competitive Lease Sale No. 821 with Reference to the Application of Accept/Reject Criteria for High Bids," by Edward Erickson.

8. Recent surveys are: R. P. McAfee and J. McMillan, "Auctions and Bidding," *Journal of Economic Literature,* June 1987; R. Engelbrecht-Wiggans, "Auctions and Bidding Models: A Survey," *Management Science,* 26:2, February 1980.

9. P. Milgrom, "A Convergence Theorem for Competitive Bidding with Differential Information," *Econometrica* 47, 3, May 1979, 679–688.

10. O. Gilley and G. Karels, "The Competitive Effect in Bonus Bidding: New Evidence," *The Bell Journal of Economics and Management,* 1983.

11. A summary and bibliography of the detailed articles is in Mead et al., *Offshore Lands.*

12. Ibid., p. 54.

13. Mead et al., *Offshore Lands.*

14. U.S. Government Accounting Office, "Early Assessment of Interior's Area-Wide Program for Leasing Offshore Lands," GAO/RCED-85-66, July 15, 1985; U.S. Government Accounting Office, "Views on Interior's Comments to GAO Reports on Leasing Offshore Lands," GAO/RCED-86-78BR, March 17, 1986.

15. In point of fact, it is quite likely that the government made a very lucrative decision by making available so many leases at a time of relatively high prices.

16. Five Year Plan, Appendix P-41 through P-43.

17. Government Accounting Office, "Early Assessment."

18. A technical note: GAO's primary estimator was least-squares using their assumed recursive equation structure. They also estimated the system using two-stage least-squares. In the latter case, the system was not identified.

19. If Y_1 is not in equation 2 then the denominator in equation 3 is one. The coefficient for each X is still a composite of the coefficients of both equations.

20. More detail is provided in S. Farrow, "Does Areawide Leasing Decrease Bonus Revenues," *Resources Policy,* December 1987. Table 5-1 is reprinted by permission from that source.

21. Collinearity between the measured percent of joint bids and the price may explain the insignificance of the price variable in the majority of the high bid regressions estimated by the GAO.

22. This is particularly true if one wishes to recover the coefficients of the underlying industry equation and not the coefficients of the equation using the measured variables.

23. This is a policy proposal to include a means for state and interest groups to defer leases at the actual leasing stage. This proposal is not a current part of OCS management. Further

details are provided in S. Farrow, "Lease Delay Rights: Market Valued Permits and Offshore Leasing," *Resources Policy,* June 1987. Table 5-2 is reprinted by permission from that source.

24. The concept of lease delay rights is closely related to the concept of marketable permits for emissions trading. That literature has developed a variety of alternative means of allocating prior rights. See, e.g., R. Hahn, *Primer on Environmental Policy Design,* (Chur: Harwood Academic Publishers, 1989); and T. Tietenberg, *Emissions Trading* (Baltimore: Johns Hopkins Press, 1984).

25. CFR section 256.35(b).

26. CFR section 256.47(b).

27. The analysis in the text assumes that the lease delay cost is positive. The present value of delayed income also depends on the expected value of the lease at the time of the next sale and the probability of a bid for the lease. Throughout this analysis it is assumed that if a bid is received in the current time period, a bid will be received at the next lease sale.

28. CFR 256.46.

29. The delay bids of the sigroups are added together because of the simultaneous benefits received by all of these groups. The free rider problem associated with simultaneous benefits is discussed in the next section on the interests of sigroups in bidding for delay rights.

30. Altering the outcome need not be the granting of lease delay rights, merely that development bids are increased sufficiently to potentially compensate for externalities incurred.

31. It may be that equality of return is not possible given the budget of sigroups. This can lead to sigroups pursuing only one alternative, probably nonmarket activities.

32. Further information on these calculations and the distribution of the lease delay cost can be found in Farrow, "Lease Delay Rights."

Chapter 6

1. Lease sales are sometimes canceled. Five sales were canceled in 1986, the largest number in a single year to date. An incentive for the Department of the Interior to offer at least one sale in each area during the span of the five-year plan is that no sales may be added to the plan, only deleted.

2. More detail exists in the court decisions 668 F.2d 1290 (1981) and 712 F.2d 584 (1983) and in the five-year plan. For a critical review of these decisions, see E. Fitzgerald, "California v. Watt: Congressional intent bows to judicial restraint," *Harvard Environmental Law Review,* 11:147, 1987, 147–201.

3. See, e.g., R. Zeckhauser and E. Stokey, *A Primer for Policy Analysis* (New York: Norton, 1978); D. Bohi and M. Toman, *Analyzing Nonrenewable Resource Supply* (Baltimore: Resources for the Future, 1984); P. Dasgupta and G. Heal, *Economic Theory and Exhaustible Resources* (Cambridge: Cambridge University Press, 1979); M. Weitzman, "Optimal Search for the Best Alternative," *Econometrica,* May, 1979.

4. S. Farrow, "Testing the Efficiency of Extraction from a Stock Resource," *Journal of Political Economy,* June 1985; R. Halvorsen and T. Smith, "Extraction in the Canadian Metal Mining Sector," *Journal of Political Economy,* March 1985; M. Miller and C. Upton, "The Hotelling Valuation Principle," *Journal of Political Economy,* 1984.

5. T. Teisburg, "Federal Management of Energy and Mineral Resources on the Public Lands," *Bell Journal of Economics and Management,* 39:2 (1980), 448–465; M. Rose and D. Rosenthal, "The Timing of Oil and Gas Leasing on the Outer Continental Shelf: Theory and Policies," *The Energy Journal,* April, 1989.

6. R. Gilbert and C. Mason, "Reducing Uranium Resource Uncertainty: Is it Worth the Cost?" *Resources and Energy* 3:13–37, 1981.

7. J. Swierzbinski and R. Mendelsohn, "Information and Exhaustible Resources," *Journal of Environmental Economics and Management,* May, 1989; S. Farrow and J. Krautkraemer, "Extraction at the Intensive Margin: Metal Supply and Grade Selection Response to Anticipated and Unanticipated Price Changes," *Resources and Energy,* March, 1989.

8. R. Pindyck, "The Optimal Exploration and Production of Nonrenewable Resources," *Journal of Political Economy,* 5:841–861, 1978.

9. M. Freeman, "The Quasi-Option Value of Irreversible Development," *Journal of Environmental Economics and Management,* September, 1984.

10. P. Dasgupta and J. Stiglitz, "Resource Depletion Under Technological Uncertainty," *Econometrica,* January 1981; W. Nordhaus, "Production with a Backstop Technology," *Brooking Papers on Economic Activity,* 1979.

11. P. Dasgupta, R. Gilbert and J. Stiglitz, "Strategic Considerations in Invention and Innovation: The Case of Natural Resources," *Econometrica,* Sept. 1983.

12. R. Stroup and J. Baden, *Natural Resources: Bureaucratic Myths and Environmental Management,* Pacific Institute for Public Policy Research, 1983; S. Brubaker, ed., *Rethinking the Federal Lands* (Baltimore: Resources for the Future, 1984).

13. Research has indicated that companies do consider these factors in their bids for leases. See T. Grigalunas and J. Opaluch, "Controlling Stochastic Pollution Events through Liability Rules: Some Evidence from OCS Leasing," *The Rand Journal of Economics,* Winter 1984, pp. 142–151.

14. This section is based on S. Farrow, "Financial Accounting for the OCS Leasing Program," *Oceans 89,* Vol. 2, IEEE Publication 89CH2780-5.

15. *Report of the President's Commission on Americans Outdoors,* Island Press, Washington, D.C., 1987.

16. D. Rosenthal, M. Rose, and L. Slaski, "The Economic Value of the Oil and Gas Resources on the Outer Continental Shelf," *Marine Resource Economics,* 1989, pp. 171–189.

17. Rosenthal et al., "Economic Value."

18. M. Boskin et al., "New Estimates of the Value of Federal Mineral Rights and Land," *American Economic Review,* December 1985.

19. This section is based on S. Farrow, "Does Analysis Matter? Economics, Litigation and Planning in the Department of the Interior," School of Urban and Public Affairs Working Paper 88-1, Carnegie Mellon University. A related analysis applied to wilderness area planning in the U.S. Forest Service is by P. Mohai, "Public Participation and Natural Resource Decision Making: The Case of the RARE II Decisions," *Natural Resources Journal,* Winter 1987, pp. 123–155.

20. D. McCloskey, "The Loss Function has been Mislaid: The Rhetoric of Significance Tests," *American Economic Review,* May 1985, pp. 201–205.

21. McCloskey, "Loss Function."

22. This differs from a multinomial choice or ordered probability model because sales are cardinal instead of ordinal numbers, i.e., the sales schedule modified by an arbitrary constant is a very different schedule. See G. S. Maddala, *Limited-Dependent and Qualitative Variables in Econometrics,* (Cambridge: Cambridge University Press, 1983).

23. See, e.g., A. Mood, F. Graybill, and D. Boes, *Introduction to the Theory of Statistics* (New York: McGraw-Hill, 1974); J. Hausman, B. Hall, and Z. Griliches, "Econometric Models for Count Data with an Application to the Patents-R&D Relationship," *Econometrica,* July 1984.

24. Significance at the 95 percent level is based on the asymptotic t ratio reported in parentheses. The small sample size, 26, indicates caution when interpreting asymptotic results.

25. The probit model is discussed in many econometrics textbooks such as J. Kmenta, *Elements of Econometrics* (New York: Macmillan, 1986).

26. A contingency table was also investigated for this sample, which failed to reject the hypothesis that the number of sales was independent of the groupings of the Department of the Interior. The sparseness of the table, however, suggested caution in its interpretation.

27. Interview with author, September 11, 1987.

28. This prediction was accurate as described in the introduction.

Chapter 7

1. J. Broadus, "Seabed Materials," *Science* 235(4791): 853–860; 1987.

2. See, e.g., U.S. Bureau of Mines, "An Economic Reconnaissance of Selected Heavy Mineral Placer Deposits in the U.S. Exclusive Economic Zone," Open-File Report OFR 4-87, 1987; Office of Technology Assessment, U.S. Congress, "Marine Minerals: Exploring Our New Ocean Frontier," 1987.

3. "United States Ocean Policy," statement by the president, 119 Weekly Compilation of Presidential Documents 383 (March 10, 1983.)

4. U.S. Office of Technology Assessment, "Strategic Materials: Technologies to Reduce U.S. Import Vulnerability," OTA-ITE-248, Washington, D.C.

5. In 1989 a slightly amended version of the former bills was introduced as HR 2440. This version included a provision allowing the administrating agency, in this case NOAA, to allocate exploration licenses through a competitive auction process.

6. 92 Int. Dec. 459, "Authority to issue OCS Mineral Leases in the Gorda Ridge Area," 1985.

7. P. Hoagland III, "The Conservation and Disposal of Ocean Hard Minerals: A Comparison of Ocean Mining Codes in the United States," *Natural Resources Journal*. 28(3):451–508, 1988.

8. Bureau of Mines, "An Economic Reconnaissance of Selected Sand and Gravel Deposits in the U.S. Exclusive Economic Zone," Open-File Report OFR 3-87, 1987.

9. Bureau of Mines, "Economic Reconnaissance."

10. K. Emery, and L. Noakes, "Economic Placer Deposits of the Continental Shelf," Technical Bull., United Nations, ECAFE, Vol. 1, pp. 95–111.

11. State of Hawaii Department of Planning and the Minerals Management Service, *Mining Development Scenario for Cobalt-Rich Manganese Crusts in the Exclusive Economic Zones of the Hawaiian Archipelago and Johnston Island*, 1987; and by the same authors, *Proposed Marine Mineral Lease Sale in the Hawaiian Archipelago and Johnston Island Exclusive Economic Zones*, 1987.

12. J. Broadus, "Economic Significance of Marine Poly-Metallic Sulphides," proceedings of the 2nd International Seminar on Offshore Mineral Resources, Brest, France; GERMINAL, pp. 560–576.

13. P. Hoagland III, "Seabed Mining Patent Activity: Some First Steps Toward an Understanding of Strategic Behavior," *J. of Research Management and Technology*. 14(3):211–222 (1986).

14. Minerals Management Service, "Prelease Prospecting for Marine Mining Minerals other than Oil and Gas," *Federal Register*, 52:8, March 26, 1987, pp. 9758–9766.

Chapter 8

1. The interviews took place prior to the spill from the *Exxon Valdez* and so no references are made to that event.

2. *Los Angeles Times,* September 7, 1987, p. 8.

3. George Stigler, "Comments on Jaskow and Noll," in *Studies in Public Regulation,* Gary Fromm, ed., Cambridge: (MIT Press, 1981).

4. N. Bockstael, W. Hanemann, and I. Strand, "Measuring the Benefits of Water Quality Improvements Using Recreational Demand Models," Environmental Protection Agency Cooperative Agreement Cr-81143-01-1, 1986; N. Bockstael, K. McConnell, and I. Strand, "Benefits of Improvement in Chesapeake Bay Water Quality, Environmental Protection Agency Cooperative Agreement CR-81143-01-0, 1987; U.S. Department of Commerce, NOAA Public Area Recreation Visitors Survey, Cooperative Agreement, NOAA and the U.S. Forest Service.

5. Minerals Management Service, "Review of Selected Fisheries Issues," Technical Resources, Inc., OCS Study MMS 89-004, Vienna, VA. 1989; 1989 U.S. App. Lexis 10156.

6. Summaries of these issues can be found in R. Cummings, D. Brookshire, and W. Schultze, *Valuing Environmental Goods: An Assessment of the Contingent Valuation Method* (Totowa: Rowman and Allenheld 1986); M. Freeman, *The Benefits of Environmental Improvement* (Baltimore: Resources for the Future 1979); W. Desvousqes, R. Dunford, J. Domancio, *Measuring Natural Resource Damages: An Economic Appraisal,* Research Triangle Report RTI/3981-00-FR to the American Petroleum Institute, 1989.

7. A recent paper with substantial references on this topic is by D. Epple, "Estimation of Characteristics," *Journal of Political Economy,* 1987; see also E. Wilman, "External Costs of Coastal Beach Pollution: A Hedonic Approach," Resources for the Future, 1984.

8. Minerals Management Service, "Impacts of Outer Continental Shelf Development on Recreation and Tourism: Volume 3," MMS 87-0066 pp. 69–72.

9. National Research Council, *The Adequacy of Environmental Information.* An immediate effect associated with the report was a recommendation by the review committee to postpone scheduled lease sales off California and Florida.

10. C. Camerer and H. Kunreuther, "Decision Processes for Low Probability Events: Policy Implications," *Journal of Policy Analysis and Management,* Fall, 1989.

11. C. Mason, "Exploration, Information and Regulation in an Exhaustible Mineral Industry," *Journal of Environmental Economics and Management,* 1986; R. Cairns, "The Economics of Energy and Mineral Exploration: An Interpretive Survey," mimeo, Centre for the Study of Regulated Industries, McGill University, Montreal, Quebec, Canada; applications include R. Pindyck, "The Optimal Exploration and Production of Nonrenewable Resources," *Journal of Political Economy,* 86:51, 1978; J. Livernois, "Estimates of Marginal Discovery Costs for Oil and Gas," *Canadian Journal of Economics,* May, 1988.

Appendix

1. Minerals Management Service, *Offshore ADP Activities: Status and Fiscal Year 1986 Projections.*

2. U.S. Dept. of the Interior, *Measuring Damages to Coastal and Marine Natural Resources,* PB87-142485, January, 1987. The computer program for assessing damages is a part of the report.

Index

About the Authors

R. Scott Farrow is an associate professor of economics in the School of Urban and Public Affairs at Carnegie Mellon University. During an interpersonnel agreement with the Department of the Interior, he worked in the Branch of Economic Studies, Offshore Resource Evaluation Division of the Minerals Management Service. Dr. Farrow has also been a Marine Policy Fellow at the Woods Hole Oceanographic Institution. His writings have appeared in journals such as *The Journal of Political Economy, Resources Policy,* and the *Journal of Policy Analysis and Management* among others.

James M. Broadus is the director of the Marine Policy and Ocean Management Center at The Woods Hole Oceanographic Institution. He is a member of the United Nations Joint Group of Experts on Scientific Aspects of Marine Pollution, and of the Advisory Panel to the U.S. Office of Technology Assessment on Technologies for Exploring and Developing the Exclusive Economic Zone. He is also a member of the Marine Board of the National Research Council, a part of the National Academy of Sciences. His writings have appeared in *Science, Marine Resource Economics,* and *Resources and Energy* among others.

Thomas A. Grigalunas is a professor in, and former chairman of, the Department of Resource Economics at the University of Rhode Island. He is a former chairman of the national Scientific Committee of the OCS Advisory Board of the U.S. Department of the Interior. He was an economist on the National Oil Spill Research Review Panel and coprincipal investigator of the Natural Resource Damage Assessment Model for Coastal and Marine Environments and for the analysis of the environmental costs of the most recent five-year OCS oil and gas leasing program. His articles have appeared in journals such as *The Rand Journal of Economics, The Natural Resources Journal,* and *Marine Resource Economics* among others.

Porter Hoagland III has worked as a research associate at the Marine Policy Center at the Woods Hole Oceanographic Institution and at the National Wildlife Federation. He has published papers in *The Natural Resources Journal* and in the *Journal of Resource Management and Technology* among others. He is currently in the Ph.D. program at Harvard's John F. Kennedy School of Government.

James J. Opaluch is an associate professor in the Department of Resource Economics at the University of Rhode Island. He is an associate editor of *The Journal of Environmental Economics and Management.* In 1986 Dr. Opaluch organized a national workshop on marine pollution and environmental damage assessment. He served as coprincipal investigator for the Natural Resource Damage Assessment Model for Coastal and Marine Environments and for the analysis of the environmental costs of the most recent five-year OCS oil and gas leasing program. His papers have been published in *The Journal of Environmental Economics and Management, The Rand Journal of Economics,* and *The American Journal of Agricultural Economics* among others.